Man and Machine

The Best of
Stephan Wilkinson

The Lyons Press

Guilford, Connecticutt

An imprint of the Globe Pequot Press

"Albatros," "Vern Raburn's Connie," and "Flying Fortress" were previously published in *Pilot* magazine.

"The World's Loudest Airplane," "Vee is for V12," "Fling-Wings" and "Tanks, Hot Rods, and Salt" were previously published in *Air & Space Smithsonian* magazine.

"Quarter-Pint" is an original.

All other stories were previously published in *Popular Science* magazine.

10 9 8 7 6 5 4 3 2 1

Printed in the United States of America

Library of Congress Cataloging-in-Publication Data

Wilkinson, Stephan.
  Man and machine / Stephan Wilkinson.
    p. cm.
  ISBN 1-59228-812-X
  1. Wilkinson, Stephan. 2. Automobile engineers--United States--Biography. 3. Air pilots--United States--Biography. 4. Airplanes--United States--History. 5. Motor vehicles--United States--History. 6. Private flying--United States. I. Title.
TL140.W55A3 2005b
629.222092--dc22

                        2005026373

CONTENTS

✿

EDITORS

✿

Having been a magazine writer forever, I've worked for lots of editors. A few were so bad they never should have given up their day jobs, but the vast majority have been splendid—smart, creative, talented, insightful, fun to work with and always part of that hugely rewarding endeavor that turns an idea and a string of words into a useful magazine article.

The chapters of this book grew from my relationships with three of the grandest editors I've known: James Gilbert of *Pilot* magazine, George C. Larson of *Air & Space Smithsonian* and Scott Mowbray of *Popular Science*. They were the people who assigned, published and paid for (that's important) almost all of the work that I've collected here.

James Gilbert and I grew up together as young aviation writer/photographers at *Flying* magazine, but it was James who literally took me under his wing. For he—a handsome, haughty and overbearing Brit (I thought at the time) had already won the aerobatic championship of England, and I was a baby pilot. Yet James taught me everything he could while we yowled at each other like skinny tomcats, competing for pride of place in the staff litter.

I guess I won, because irascible James got fired. But he went back to London and built his own magazine, *Pilot*, and when I eventually got fired myself, he gave me, a brand-new freelancer, lots of work to do, some of which is in this book. James had the last laugh: he recently sold his magazine to a huge British publishing company and retired rich beyond at least my wildest dreams.

It was my vote as executive editor that got George Larson hired at *Flying*, but it was his brilliance that affirmed it. Multi-talented—he went to Harvard and then Tufts Medical School, went to Vietnam as a newly minted lieutenant and worked with hill-country Montagnards, then came home to briefly become a folk-rock musician—George would have been a great doctor and a great folksinger, but fortunately, he instead became a great editor.

After paying his dues at *Flying*, Larson got hired by the countercul-ture magazine *Mother Earth News* and moved wife, kids, and furniture from LA to North Carolina to take the job. It lasted exactly one month, when the loose cannon who owned the magazine decided he didn't like George after all. Larson ultimately became the first editor of a brand-new magazine called *Air & Space*, published by the Smithsonian Institution—a novel attempt to create an aerospace magazine for people who weren't necessarily pilots, just enthusiasts.

Being the founding editor of a start-up is the kiss of death. Typically, when the new magazine is still undergoing inevitable birth pangs after a year or two, the suits all blame it on the editor and fire him or her. George Larson is to this day, 21 years later, still running *Air & Space*. Other than the late George Plimpton, I can't think of another founding editor who hung onto the desk that long. Bless him, at least in part because George has the uncanny ability to write assignment letters so detailed and artful that they are often better than the article itself.

Another writer once said of Scott Mowbray, "The man has no right to know as much about as many things as he does." Mowbray took a dying magazine—*Popular Science* in the late '90s had become a shadow of its fa-mous self, pamphlet-thick, devoid of any editorial vision and supported by the occasional erection-pill ad—and in two years turned it into the winner of a National Magazine Award in the General Excellence category, which is the magazine industry's equivalent of a best-picture Oscar.

Mowbray engaged me to write the regular column that gives this book its title, and doing "Man and Machine" articles for him was both a challenge and a delight. A challenge because he had a finely tuned bull-shit detector that would instantly register my "Gotta write something this month, might as well see if I can get this half-baked idea past Mowbray ..." and a delight because he could just as quickly see the merit of an idea that would have had a lesser editor musing, "Uh, I don't get it."

I'm not the only person who thinks Scott Mowbray is a brilliant editor. His publishing company, Time4 Media, a division of the huge Time Inc./Time Warner complex, has made him Editorial Director of all 16 of its magazines.

Stephan Wilkinson

INTRODUCTION

❋

FOR MY FOURTH BIRTHDAY, MY MOTHER made a cake with an airplane atop it. It was white sugar icing, I remember, a biplane. Strange thing to put on a birthday cake, since I barely knew what an airplane was. It was April of 1940, and the Phony War was quietly simmering in Europe, as Germany, France, and England postured. The Battle of Britain was only five months away, and we'd all learn what airplanes were.

The airplane was my mother's fascination, not mine. She had grown up on New York's Long Island not far from Roosevelt Field, from which Lindbergh left for Paris on a damp morning in May of 1927, and on the edge of the Army Air Corps base at Mitchel Field, where she often danced with pilots. (I think she still carried a bit of a torch for Johnnie Johnson, of whom I would later become aware as Colonel—later General—Leon W. Johnson, one of the leaders of the huge "Ploesti Raid," when B-24s at the limit of their range tried unsuccessfully to destroy the gasoline refineries that the Germans controlled near Bucharest, Romania.)

My father, for his part, would later tell stories of bicycling from his New Jersey home to the big grass—then—field at Teterboro, where many famous pilots flew. (Well, they were *all* famous in those days.) He remembered watching Berndt Balchen make test hops in the big Fokker trimotor that Balchen would soon pilot for Admiral Robert Byrd on the first flight over the South Pole, and lord knows who else he saw without being aware of it. Perhaps even Howard Hughes . . . Roscoe Turner and his lion cub . . . Lindy . . . Amelia . . . .

Dad was one of the original "boys at the airport fence," and it's no wonder I too became fascinated by airplanes. Kids don't ride bikes anymore, but my friend Jerry Slocum was a senior captain for Delta, who when he was based at Salt Lake City once told me, "I'll know that airplanes are no longer fascinating on the day we taxi out for takeoff and there isn't a single car parked in the turnoff at the airport fence, somebody just sitting there watching takeoffs and landings. But so far, no matter how bad the weather in winter, no matter how early in the morning our departure, it has yet to happen."

Hardly surprising that I went on to become a pilot. Because of that simple skill, I met my wife, who is also a pilot. We flew together for years, shouting at each other all the while, and not simply because the airplanes were loud.

Flying paralleled my fascination with fast cars, and each fed off the other. Just as mathematicians often are also musicians—the two talents are closely related—airplanes and automobiles attract the same motion junkies. Some pilots are enormously conservative, careful, serious. They are superb pilots, safe pilots. They drive very plain cars and don't smile much. The rest of us drive fast cars too fast and look back to wonder how we survived our aerial stupidities.

Captain Slocum, who carried millions of unsuspecting Delta passengers in perfect safety, none of them suspecting they were being levitated by an ex-street racer who at 19 was the terror of Van Nuys Boulevard in his Cadillac-V8-powered Model A Ford, is now retired. Yet Jerry recently asked me for some car-buying advice, admitting that he was attracted to the 604-horsepower Mercedes-Benz SL65 convertible supercar.

"Yeah, I drove it a while ago," I e-mailed him in response, "and it's pretty insane."

"That's the ultimate compliment," he answered. "My grandfather drove sane cars. I am not my grandfather. I am an incurable hot rod–era guy who will always be drawn to vehicles with more horsepower than any rational human being could ever possibly need."

We are fascinated by machines, people like Jerry Slocum and me. No, we don't "love" them—they are simply machines, after all, and such a powerful and meaningful emotion needn't be wasted on them. Machines do our bidding, they hugely increase our capabilities, they reflect our pride as well as our insecurities—think midlife-crisis Corvette—and they are, for better or worse, a stereotypically masculine fascination.

Perhaps this is because most women are innately too intelligent to toy with the dark side of machinery, though men can't resist going there. Certainly not all, as Indy driver Danica Patrick, three-time U.S. National Aerobatic Champion Patty Wagstaff, and any number of Air Force, Navy, and Marine women fighter and helicopter pilots amply demonstrate. But fast cars, speedboats, superlight bicycles, motorcycles that put the power of a compact car in your crotch, .50-caliber sniper rifles the size of a Spitfire's wing gun—they all shake the male cocktail of testosterone with a dash of adrenaline.

It has nothing to do with bravery. I am a professional coward and would no more seek physical confrontation than I would stick my hand in a Cuisinart, but I do love the jolt that comes from putting myself into the grip of a machine that can do me harm. Without that infrequent but very real element of danger, my life would have been vastly less stimulating, vastly more ordinary. Boys and their toys . . . the machines we play with are what separate us from lesser beings, fill our days with adventure and keep at least some of us young. ✿

The World's Loudest Airplane

※

It was the "Sounded Like a Good Idea at the Time" era. A time when more outlandish, imaginative, free-thinking—and in some cases, totally goofy—aeronautical concepts were built and flown than at any time before or since. Airplanes that took off straight up, hanging from enormous contra-rotating props, or climbing the beanstalk of their jet thrust. Tiny jet fighters that joined up with the bellies of bombers like suckling kittens, hanging from hooks that towed them along. Jets that took off from trucks, flung into the air by rockets. Inflatable airplanes. Flying wings. Tailless deltas. Jet seaplanes. Jet seaplane fighters, forgodsake.

So how's this one sound? Let's take an early but perfectly adequate jet fighter, the Republic F-84—in fact, our first axial-flow jet fighter, with a bud vase of a fuselage that took advantage of the tubular slimness of the axial-flow engine—and put a propeller on it.

But isn't aviation trying to get rid of propellers?

Never mind, we're going to drive this propeller with an enormous turboprop engine—two engines, in fact, coupled together through a common gearbox—and we'll spin it so fast that the prop tips will be traveling at almost 1,260 mph—Mach 1.71. At least the prop will be supersonic, even if the airplane isn't.

The result, in 1955, was the Republic XF-84H, a swept-wing, single-seat, T-tail turboprop that had the unhappy distinction of being the loudest airplane ever built.

The original F-84 was the Thunderjet, to remind everyone that it was part of a family that began with the World War II P-47 Thunderbolt. Its swept-wing follow-on, the F-84F, was the Thunderstreak, which was followed by a reconnaissance version, the RF-84F, called the Thunderflash.

The XF-84H, however, was ingloriously dubbed the Thunderscreech.

"One day, the crew took it out to an isolated test area [at Edwards Air Force Base, in California] to run it up," recalls Henry Beaird, a Republic test pilot who was one of only two men ever to fly the Thunderscreech. "They tied it down on a taxiway next to what they assumed was an empty C-47, but that airplane's crew chief was inside, sweeping it out. Well, they cranked that -84H up, made about a 30-minute run, and shut it down. As they were getting ready to tow it back to the ramp, they heard this banging in the back of the C-47. Turned out the noise had made the guy lose control of his arms and legs, and he was just flailing around, laying on the floor. He eventually came out of it."

Says Beaird, today in his 80s and still flying Learjets, "As long as you stood ahead of or behind the airplane, it really wasn't so bad, but if you got in the plane of the prop, it'd knock you down." Really? "Really."

But there was a good reason to build the Thunderscreech. Early jets—P-80s, F-84s, even the vaunted F-86—were like powerful but overgeared vintage Ferraris. Put the thing in fourth and step on it and you may eventually do 150, but you'll be forever getting there. The jets accelerated, lifted off, and initially climbed achingly slowly. This meant that either they needed long runways if they were loaded for bear—a fighter's natural stance—or they were limited in how much fuel or weapons they could carry. And on landing, a turbojet pilot had to be very careful about speed control: get a little too slow on short final and you'd hit the ground before the jet engine woke up and put out enough thrust to accelerate.

Propellers were different. The most powerful of them had the opposite problem: You'd better feed in power judiciously, for they reacted so violently to a firewalled throttle that the entire airplane tried to counterrotate against the prop's torque. With a tractor propeller spinning clockwise (as seen from the cockpit), the effect was that the airplane would turn hard left and plow straight off the runway. But if nothing else, they provided power right now.

So the Air Force's Propeller Laboratory, at Wright-Patterson Air Force Base in Dayton, Ohio, continued to experiment with propellers, and it was for them that the XF-84H was built as a testbed, to examine the efficacy of supersonic propellers. "That didn't mean the airplane would run supersonic," Beaird warns, "because with that big a prop disc up front, it's like a big speed brake. It meant that on the -84H, the outer 12 to 18 inches of the propeller were supersonic all the time."

That, of course, was the source of the horrendous noise. Nor was there any way to mitigate it, for the Thunderscreech's engine ran at full speed all the time, and the propeller rotated at 3,000 rpm from start-up till shutdown. "All you had to do was move the propeller pitch control to get power," Beaird explains, "and you got it pretty instantaneously. Maybe it got a little louder when you added power, because I do remember hearing it better, 22 miles away from the base where I lived, when they'd run it up to full power.

"Edwards was worried that the noise of the airplane would break the windows in the control tower," Hank Beaird remembers. "The runway's about a mile from the tower, but they'd put blankets over the top of the shelf where the radios were, and they'd get up under their desks, under the blankets. Nobody ever actually recorded the decibels. I think they were afraid the measuring device might get broken."

Edward von Wolffersdorff was Beaird's crew chief. "Oh, man, that noise was terrible. You can't imagine," Ed groans. "I remember making my first ground runs with the thing, down on the main base, and I was wondering, why are they flashing that red light at me over on the control tower? It turned out they couldn't hear a damn thing over their radios, so they kicked us out and sent us over to the north base."

The airplane was not popular at Edwards and is to this day rumored to have caused several miscarriages. "It's hard working on a project like that when you know everybody's against it," von Wolffersdorff admits.

"We were trying hard to get this thing going, and we didn't get the support we needed. Nobody wanted the damn thing. First the Navy backed out, and then the Air Force canceled the project. A lot of people thought we were trying to go supersonic with a prop, but that wasn't true at all."

The Navy had gotten wind of the -84H and initially wanted in on the project, so three were scheduled to be built by Republic—two for the Air Force and one for the Navy. (In the end, only the first two made it out the door.) The Navy liked the fast-turboprop concept because pure jets were problematic aboard carriers. The catapults of the era had a hard time accelerating fighters to takeoff speed. And on landing, standard procedure was to go to full power right at touchdown, to carry the airplane back off the deck if the tailhook missed the arresting wires. Jets were slow to respond in a "bolter."

Three manufacturers were asked to provide experimental props for the -84H—Aeroproducts, Curtiss-Wright, and Hamilton Standard. In the end, only Aeroproducts stepped up to the plate, providing a stubby three-blade paddle prop, the blades only about four times as long as each was wide. "It was a funny-looking propeller," Beaird recalls. "I think it was just one they happened to have available in the right diameter."

The Thunderscreech's Allison T40 engine was—even in the words of the company's authorized history, *Power of Excellence*—"a monstrosity, a mechanical nightmare. . . . Allison was in the throes of developing the turboprop concept, and began probably 20 years ahead of where it should have been." The T40 was a pair of 2,750-shaft-horsepower T38s inside a common case and mounted behind the cockpit, where the F-84's General Electric J35 turbojet had originally lived. (Although the -84H's swept wings and main landing gear were straight off the RF-84F, the Thunderscreech's fuselage was almost entirely new, substantially modified to fit the big T40 engine. In fact, the airplane was different enough from the basic F-84 that it was originally to be called the XF-106, a designation that eventually was given to the Convair Delta Dart interceptor, which went on to serve in Vietnam.)

At the time—the mid-1950s—the T40 was the most powerful aircraft engine on the planet, putting out between 5,850 and 7,400 shp, depending on the model. Each of its T38s turned an 18-foot-long driveshaft that led to a big gearbox in the XF-84H's nose, directly behind the prop. Though the shafts weren't visible to the pilot, they were spinning at stunning speed on either side of the cockpit, just under the floorboards. One of Republic's major concerns was that the driveshafts would overheat the numerous bearing blocks through which they ran, in an attempt to stiffen the relatively flexible shafts as much as possible. Each bearing had temperature and vibration sensors, with both meters and warning-light readouts on the glareshield directly in front of the pilot.

"We looked at the damn gearbox and thought, jeez, that's gonna be a bear," Ed von Wolffersdorff recalls. "And those shafts that ran past the cockpit on each side—boy, that made you pucker up just to think about it. We were expecting the worst, but they never gave us a bit of trouble.

"We did have some problems with the gearbox, but it was operator error. You'd get the left engine going first, then you'd engage its clutch and get the gearbox turning, drive the right-hand engine back through the gearbox, and get it going . . . . I was checking out another crew chief and told him to be careful, but he forget to get the coolant oil flowing and man, it just cooked one clutch."

The entire starting procedure consumed half an hour, Hank Beaird says—building up hydraulic pressures, establishing good electric-power levels, getting the proper green lights, moving from step to step in the sequence.

"In flight, the driveshaft vibrations were usually up in the high levels," Beaird recalls. "It was very sensitive. If it got to where the vibration was so bad that I thought it was going to cause damage, they just left it up to me to decide whether to get out of the airplane." In testpilotspeak, "get out of the airplane" means eject. Beaird never did that, but 10 of his 11 Thunderscreech flights ended in premature or emergency landings due to vibration or prop-controller problems. "The only time it became a good handful was when you got it out around 400 knots," he says. "The propeller governor would start surging, and the airplane would roll rather violently." The entire airframe was trying to rotate around the propshaft, torquing like a big flywheel with wings.

The late Lin Hendrix—another Republic test pilot who made a single Thunderscreech flight and was the only pilot to fly the second of the two airplanes—once wrote that Beaird, "who never swore, once said after an emergency landing, 'By jingo, that airplane is going to hurt somebody!'" Hendrix himself declined any further opportunities to fly the 'Screech, telling Republic's chief engineer, a muscular 6'4" and 235 pounds, "You aren't big enough and there aren't enough of you to get me in that thing again."

Only a single XF-84H survives, that number-two airplane having been junked. The original testbed spent several decades on a plinth at the entrance to the Bakersfield, California, Municipal Airport, where an electric motor in the spinner turned the prop at a stately 10 rpm, hardly hinting at the 'Screech's bad habits. The old gate guardian slowly baked in the high-desert sun until 1992, when it was given hangar space by the USAF Museum, in Dayton, Ohio. It has since been restored to display condition, and less than a year ago was finally put on exhibit in the Museum's experimental-aircraft hangar.

Robert Schneider and Darrell Larkin had both flown F-84s in the Ohio Air National Guard, and they assembled a team of volunteers who spent a total of 3,710 man-hours doing that restoration. "You know you're in trouble when you have to have pilots working on an airplane," Schneider laughs, "but Darrell and I found a lot of retired chief master sergeants who'd been sheet metal guys and had other specialties. They're the ones who really did the work."

Why did they use F-84s as the testbed for the supersonic-prop experiment, as well as a considerable variety of other oddball projects? "It had a roomier cockpit than the F-86, and there were a lot of them made," Bob Schneider says. "It was a good-flying aircraft—a little underpowered but extremely strong. I had a midair collision once with another F-84, and we both kept flying and landed safely." Schneider and a flight of three other F-84s were vectored into a thunderstorm by a careless ground controller, and in the murk, the -84 to his right slammed into Schneider's airplane, its stabilator shearing off the front of his wingtank and then whacking the fuselage.

"I barely knew what had happened, other than a big bang and a shadow crossing over my canopy, but I checked the fuel and there were 1,500 pounds in the left wingtank and zero in the right one. I looked out and saw that there essentially was no right wingtank. When the other guy landed, his left stabilator was bent down at a 45-degree angle."

"The XF-84H was a hulk when we got it," Darrell Larkin recalls. "I think every kid who ever walked by that airplane in Bakersfield threw a rock up the tailpipe. I had to take a ton of stones out of there." But the airplane had never been vandalized, since it was on airport property and reasonably secure. "Except for the birds and other animals, prairie dogs, I don't know what," Larkin says. "There were nests everywhere. We had to do a lot of vacuuming, clean up a lot of dirt."

Did the Air Force appreciate their efforts? Yes . . . and no. "We were going to restore an RF-86 after we finished the -84H," Larkin muses. "The airplane had been through two floods, and it needed a lot of work. There was no dash-one on the R version [what the Air Force calls a dash-one is an aircraft's official operations-and-maintenance manual], but I found a private source for one and suggested to the Museum that we buy it. I got a full-page, single-spaced letter of reprimand from the Museum for not going through the proper channels. I told 'em that was it, I didn't want to have anything else to do with the Museum."

Though its stubby but strident propeller got all the attention, the XF-84H was a precedent-setting airplane in other ways as well. It ultimately proved to be a dead end, but the 'Screech was—and still is—the world's only turboprop with an afterburner, and visitors to the Air Force Museum can peek into the tailpipe and see all the spraybars and plumbing still in place.

Turboprops typically use their engine's tailpipe simply as a vent for spent gases that have already done most of their work, though the exhaust flow usually produces residual thrust as well—almost 1,300 pounds' worth, in the case of the Thunderscreech. The Navy wanted all the carrier-take-off thrust they could get, so Allison fitted the baby burner to the T40 at the USN's request. It was only lit off on the test stand, never in flight.

The -84H was also the first airplane to carry a "rat"—a ram-air turbine that automatically deployed from a niche inside the airplane's dorsal fin and pinwheeled in the airstream to provide extra electrical and hydraulic power.

"The airplane had full-span ailerons whenever the gear was down, since the flaps became ailerons too," Hank Beaird explains. "It took a lot of power to move those surfaces if you had to move them in a hurry, and the rat provided that. It would come out whenever the gear was down. That was one of the airplane's biggest contributions. We put that on other jets as well—particularly the F-105, which also had the full-span aileron system."

Another XF-84H feature that Beaird liked was its speed brakes, all the way aft alongside the afterburner nozzle, that opened to each side like flower petals. "I don't think anybody had ever put them as far back as we had them on the -84H," he says. "Yeah, we learned a few things with that airplane. We put the same speed brakes on the F-105, but bigger—a four-pedal arrangement. They made little or no trim change but tremendous drag. On the -105, you could put those things out at 1.8, 1.9 Mach and you'd just be standing on the rudder pedals, it slowed down so fast."

The two XF-84Hs flew a total of less than 10 hours, all of it at the hands of two civilians, Beaird and Hendrix. It remains the only aircraft bought and paid for by the U.S. Air Force that has never been flown by a military pilot. (The Navy bailed out before their part of the bill came due.) Nobody knows how fast a production F-84H might have gone. Republic, wildly optimistic, foresaw 670 mph, a range of over 1,000 miles, and a rate of climb that would have taken it to 35,000 feet in 12 minutes. Neither of the two X-planes ever made it past 450 mph.

Still, this was thought fast enough at the time to make the XF-84H the fastest prop-driven airplane in the world, a claim that occasionally is still repeated. In fact, that speed record was already held by the huge four-engine, eight-prop Soviet Tupolev Tu-95 Bear bomber, which, able to cruise at 545 mph, remains by far the world's fastest propeller aircraft. The turboprop Bear was already in service in 1955, when the XF-84H made its first flight. When the Bear first appeared, Western "experts" pegged its speed at 400 mph, based on what they knew of the XF-84H experience.

Tupolev, however, had realized that the key to high prop-driven speed was long, multiple, slow-turning blades, contra-rotating for maximum efficiency, not a screeching little three-blade paddle.

Well, it sounded like a good idea at the time. ✸

Do the Locomotion

✿

THEY SIT ON A SPUR OF TEST track outside General Electric's locomotive factory in Erie, Pennsylvania, panting and grumbling like two old lions half asleep. The ominous, muttering rumble is the idle of 8,800 horsepower—24 cylinders with pistons big as buckets, turbochargers the size of washing machines, two V12 engines driving alternators five feet in diameter. For here are two units of the most advanced diesel-electric locomotive in the world: a pair of GE Evolutions, running all-new powerplants designed specifically to meet stiff new Tier II EPA locomotive emissions regulations that went into effect in June of 2005.

You didn't know locomotives had emissions regs? Neither did I. I assumed that the 207-ton iron gorilla of the wheeled world damn well did whatever it wished. But the new Tier II standards require substantial cuts in NOx and particulate matter, and GE, the world's major locomotive manufacturer, designed a new engine with an air-to-air turbocharger intercooler that lowers induction-air temperatures to only a few degrees above ambient, for cleaner emissions and more power, to meet them handily. Not only is the new GEVO-12 four-stroke diesel 60 percent cleaner than its predecessor, it's 3 percent more fuel-efficient as well.

That may not sound like much, but it's in fact huge; even a half-percent improvement is a big competitive advantage in loco sales. (An Evolution sells for almost $2 million.) A locomotive typically burns about 300,000 gallons of fuel a year, and saving 9,000 gallons per engine can make a big bottom-line difference.

If you want to make the savings sound huge, here's how. There are six major railroads in the U.S., and between them all, they burn 3.5 billion gallons of diesel a year. At $1.74 per gallon—a railroad's typical cost in the aftermath of Hurricane Katrina—that's more than a $6 billion fuel bill. A 3-percent fleet-wide fuel improvement would be worth about $183 million to America's railroads.

When I was a boy, my father commuted to work in New York City from Harmon, a station on the Hudson near our home. It was the last stop for all the steam engines that brought passenger trains from as far away as Chicago, for smoky steam engines had been banned from the 50-block-long tunnel under Park Avenue that led to Grand Central Station. At Harmon, the steamers were uncoupled from their trains, and odor-free electric locomotives that got their power from an electrified third rail took over.

One of my great thrills, when I got to go to Harmon—maybe Dad's war-surplus Willys Jeep beater was in the shop and Mom and I had to go pick him up—was to stand on the pedestrian bridge over the tracks while a steam engine passed just below on its way to the yard. It was a naked view of a locomotive that few people ever got, but even better was when it passed right under my feet and the column of superhot steam pounded the bridge and shook the soles of my shoes. The 1950 equivalent of parking at the airport fence right under short final and letting the landing jets part your hair.

Gone are the steam engines, of course, and most of the Toonerville Trolley electrics. The diesel blew them off the tracks. Steam engines are inefficient, even though the biggest ones, 385-ton, 4-8-8-4 wheel-group monsters, could crank out almost 7,000 hp. The best of the steamers turned 12 percent of their combustion heat into usable energy. A modern automobile engine is about 20-percent thermally efficient. And a good diesel can be 35-percent efficient (which is why diesel automobiles get better fuel mileage).

A modern diesel road locomotive is actually a hybrid. The diesel doesn't drive the train, it cranks an alternator, which powers the six huge electric traction motors that actually turn the locomotive's wheels. (One of the biggest contributors to railroad-diesel technology was the development work done during World War II on diesel engines for submarines.) Each motor is set transversely between a pair of drive wheels. On an Evo, the electric motors will put out a total of almost 60,000 foot-pounds of torque at start-up—the equivalent of about 120 Ferrari Enzos—good for a zero-to-60 time, unloaded, of just shy of 65 seconds. Rather longer, though, when dragging a typical 18,000-trailing-ton, two-mile-long coal train.

Locomotive manufacturers periodically engage in horsepower races, the old "mine's bigger than yours" business. At this point, it has become apparent that something like the Evolution's 4,400 horsepower hits a sweet spot, even though there are 6,000-hp engines out there. But 6,000-hp engines are pretty specialized, best at hauling high-speed priority intermodal cargo. If a railroad company buys a bunch of 6Ks, they have engines too buff for all the trains that need a relatively standard 4,000-hp max. And if they have a drag that requires 8,000 hp, all they can do is couple a pair of sixes and waste a third of the power.

The diesel's electric traction motors also help to brake the train. When the driver wants to slow down, he turns the motors into generators, reversing the field so the windings and armatures are making electricity rather than consuming it, and are thereby magnetically resisting the turning of the wheels. This is a lot cheaper than buying brake shoes, which don't last long trying to hold back a train as heavy as a tramp freighter.

The excess current produced is dissipated by a big, fan-cooled "dynamic-brake grid," effectively the world's largest hair dryer, near the top of the carbody, back aft. "Does the grid actually glow?" I ask GE Lead Systems Engineer Mike Schell. "It does when it catches fire," he says with a straight face. But even with the blowers at work, you wouldn't be able to enter the compartment where it lives.

A hybrid automobile would save rather than waste all this electricity, but locomotives don't have big storage batteries to make use of regenerative braking. Apparently, installing such a system isn't worth the extra cost and complexity, since it's hard to imagine that GE hasn't at least considered it.

One of the strong points of a good diesel locomotive is "adhesion," the railroad equivalent of a sports car's grip on a skidpad. Those of us old enough to have stood on the bridge at Harmon can remember the sound of a steam locomotive starting from a standstill with a big drag behind. Whoof. Whoof. Whoof . . . . *Whoofwhoofwhoofwhoof* as the steel wheels broke loose and spun, the result of more torque than traction. Since locomotive manufacturers don't have the option of making grippier tires, they have developed sensitive wheel-speed sensors—exactly like what's used in an automobile's ABS or traction-control system—that will dispense sand into the wheel-to-track interface just as was done in the steam era, but now computer-controlled.

The Evos are dual-purpose road locomotives. (There are three basic categories of locomotives: switchers, used only in railroad yards to build trains; switcher-gatherers, which have limited distance capability; and the big boys, road locos, both freight and passenger.) Evos can do "drag runs," which are heavy, single-cargo loads like ore, coal, or grain, or they can make high-speed "intermodal runs" with flatcars carrying ro-ro [roll-on/roll-off] trailers and shipping containers. Intermodal loads are full of a lot of "air" as well as product, so they're substantially lighter. It's not easy to design a machine to do both jobs efficiently. Think of this locomotive as being an athlete that is at the same time a weight lifter and a marathoner.

How much fun is it to drive a locomotive? Not much. Take a hint from the fact that every two minutes (every 30 seconds when the train is traveling faster than 50), a big red caution light that reads "alerter reset" glows on the driver's panel. The driver has 25 seconds in which to slap a yellow switch to affirm that he is indeed present and accounted for. If he doesn't, the power automatically backs down and then the brakes come on—hard.

Engineers are increasingly being redubbed drivers, much to the chagrin of traditionalists who say, "Driver? Okay, show me the steering wheel." And brakemen—the second person in the two-seat cab—are now called helpers.

When engineers sat in the very nose of the train rather than farther back in what's today called a safety cab, they'd occasionally experience a mesmerizing vertigo brought on by the rush of passing crossties under the cowcatcher, so I suppose there's precedent for this modern equivalent of a deadman's throttle. (Lightplane pilots call the phenomenon "flicker vertigo," and it can be caused by looking toward the sun or a bright light through the strobe effect of idling propeller blades.)

It can take half a mile to panic-stop a loaded train. And if locomotive brake pads are expensive, replacing tracks scalloped by sliding steel wheels is what really gets the accounting department's attention. The emergencies typically occur at grade crossings, literally every day somewhere in the country. The biggest danger a loco crew faces, up there on the square end, is not the high-speed Casey Jones dying-with-his-hand-on-the-throttle crash but the drunk in the pickup truck trying to weave through the crossing gates at three in the morning. "People think trains can stop like cars," says GE Product-Line Manager Peter Lawson, "and that's part of the issue." Another part is that noise regulations are making trains quieter, and an increasing number of municipalities are forbidding locomotive horn-blowing, particularly late at night.

Modern locomotives are heavily computerized, and the driver of an Evolution sits in a kind of glass cockpit, behind two large CRT monitors upon which he can call up some 30 different graphic pages of instruments, gauges, graphs, and information, with a separate monitor for the helper. Every aspect of the engine's health can be tracked, and GE also monitors most of its locomotives wherever they are in North America, via GPS and an OnStar-like link to the Erie factory. The telemetry will spot a fault and transmit instructions for the crew to stop at the next opportunity for maintenance. Then it'll tell the technicians what the possible causes for the fault are, and what GE thinks the most likely cause is.

The toughest duty for a crew is not a zillion-ton coal drag two miles long. On an assignment like that, you just thunder on down the main line, and there's nothing much to do. No, the worst kind of trip features a lot of ungated crossings as well as car exchanges that require the helper to constantly climb in and out of the cab.

What's the worst thing a driver can do to hurt one of these big engines? "Just sitting there driving," says Schell, "there's nothing he can do to hurt it. We've got enough protection and warnings in place to protect the temps and pressures, the cooling water, the oil, everything. The only people who can hurt an engine are the railroads, by not doing the proper maintenance."

Still, humans are ultimately in control, and they can lose control. GE's short test track in Erie ends "in a pile of dirt and a nice old lady's yard," says Pete Lawson. "Which we've needed to landscape a couple of times." ✿

Hot Date with a Handgun

✿

I HATE GUNS.

Actually, that's not true. I love guns.

Well, that's not true either. My feelings about guns are too complex and intimate to label so casually.

On the one hand, I see guns as vicious devices designed solely to propel small, hard projectiles into flesh and organs, to maim, mutilate, and destroy the beautiful human machine. On the other, they themselves are beautiful machines that fascinate me in their preciseness and purposefulness, their craftsmanship and detail, their—at least, for all of my adult life—forbidden and illicit status. Show me a gun museum and I'll sneak in as though I was going to an X-rated movie. Stand me next to a policeman on a street corner and I'll stare at the Glock in his holster as though I were peeking down Britney Spears's blouse.

Is there a sexual element to this? I have a friend, Peter Miller, a reassuringly randy bachelor cop and the chief of police in the town next to ours, who says that if there is a turn-on, it isn't women who find guns aphrodisiacal. "At least it's never worked for me," he laughs. "Women hug you, feel the lump, and go, 'Is that a gun? Eeeuw! Get it out of here!'"

Rookie cops love to wear their off-duty pistols wherever they go, but that gets old, and the veterans typically lock their piece in a gun safe as soon as they get home. For the new guys, a gun is like the stethoscope draped around the intern's neck: it establishes his place with the big boys. My youngest brother, Alec Wilkinson, was a small-town cop in Massachusetts for a year, and he remembers that his cheap .38 "made a sound like those firecrackers called ladyfingers, and everyone else on the force had pistols that made deep and bullying reports."

Alec was a bit of a buffoon as a cop, but he wrote a wonderful book about the experience, *Midnights*, and I actually once met a policeman who said he'd decided to pursue the profession after reading Alec's book as a teenager. My favorite handgun story in *Midnights* involved Alec stopping early in the morning at an all-night diner to use the john. Five or ten miles on down the road in his cruiser some minutes later, he realized that he was uncommonly comfortable. Then it hit him. He'd left his heavy, cumbersome pistol belt, pistol, cartridges, handcuffs, and god knows what-all sitting on top of the toilet tank in the diner.

Miller is a far better lawman than Alec—then an out-of-work musician—would ever have been, but even Pete had his moments. Like the time he shot his locker. "It's the only time I've ever had a gun go off accidentally," he recalled, "and to this day, I don't know what happened. I was a rookie, reloading my .38 in the locker room of the station, and boom, it blew a hole right through my locker and into the locker next to it, which unfortunately was the sergeant's. The bullet whizzed around and around inside the locker and shredded his raincoat and two uniform shirts. He never forgave me, but it taught me forever the potential of a simple pistol."

When I was a boy, my father had an ancient single-shot .22 rifle, and I remember him taking it by the barrel and flinging it as far as he could out over the waters of Croton Reservoir. This followed an incident the day before, when I had taken one of the little rimfire cartridges and, attempting to extract the powder for god only knows what purpose, had whacked the shell with a hammer to squeeze the slug out. With an odd pop, the bullet passed neatly between my spindly legs, barely missed dejeweling me, clanged off the furnace, and bumblebeed briefly around the cellar.

That was my last physical contact with guns until recently, when, ruminating over the allure these machines still had, I called Miller and asked him, "Peter, what would you say to my coming out to the range and shooting your pistol, if I bought the ammo?" Pete thought that would be great fun.

I'd expected what you see in the movies—a basement range with lots of guys wearing hearing protectors and yellow aviator glasses, man-size targets trundling back and forth on clotheslines, assumedly located in one of our rural county's two small cities.

Not quite. Past a padlocked gate and down a long, sandy, two-ruts road, in the middle of a swampy noise-buffer zone for a big nearby airport was the deserted "range." It was a clearing in the woods with a big earthen bank at one end, nothing more.

The range was under the approach path to Runway 09 at the airport. "You're not allowed to shoot at the airplanes," Pete cautioned, just as I was wondering if anyone had been tempted to point a barrel skyward. The ground was littered with brass cartridge cases and lipstick-red riot-gun shells of every size, and the berm behind the target line was probably sinking into the earth under the weight of the lead it had absorbed.

Peter had brought his own guns and two pistols for me—a .40-caliber Glock semi-automatic and a classic Smith and Wesson .38 revolver. We stapled targets to a pair of hanging Masonite sheets. "These are called bottle targets," he said, and indeed they were: featureless, body-size white jugs the shape of Russian nesting dolls on gray backgrounds. Apparently the realistic silhouette-of-a-man targets that Dirty Harry practiced with had been declared politically incorrect, at least in part because they invariably were black.

The Glock had a malevolent feel from the minute I slapped the 15-round magazine into the butt, imitating Pete's practiced move. I held it as though it were a dangerous snake, and remembered my wife saying that when she was in college, a deputy-sheriff boyfriend had once let her fire his pistol out in the Vermont woods. "I shot it just once and then said okay, that's enough," she recalls. "I didn't want to hold it anymore. I thought it was going to somehow turn around and shoot me."

"Where's the safety?" I asked Peter as I next palmed the .38.

"There is none," he said. "That's a law-enforcement gun, and it's meant to be ready the instant you need it."

My hand shook slightly as I fingered the trigger—inhale, breathe out slowly, squeeze gently but steadily; the gun went off with a clap, a roar, an expanding BOOM that Hollywood hadn't prepared me for, since nobody would willingly sit in a theater amid the assault of such a sharp and visceral noise. The pistol bucked in my hand, hammering the serrations of the grip into my palm.

Surprisingly, there was a hole not far from my bottle's . . . well, its right kidney, I suppose.

I tried the standard police "double-tap" move that Pete had taught me, though I left out the fast-draw part for fear of blowing off my toes. Stand in a flat-footed crouch, raise your piece, don't bother to actually aim, and fire twice, BAMBAM, at "main body-mass." Then direct the gun down toward the ground at a 45-degree angle ready to reshoot.

Wow. Two more holes.

"Here's what I like to do," Peter demonstrated. "I love the noise, the violence of putting out a whole magazine." BAMBAMBAMBAMBAM the Glock went, 15 times, several of the warm cartridges bouncing off my cheek as I watched, stunned.

Well, I can do that. And I did, even more stunned by the power that erupted from my own puny hands.

Power. That's what it's about. We're a culture besotted by power. While much of the rest of the world seeks efficiency, size matters for us. We love the 500-hp Dodge Viper and find the Ford Excursion too small when a Hummer heaves into sight. We gripe about the gall of those rotten Enron execs, yet actually wish we could manipulate billions. We dimly remember when it was Charles Atlas who empowered the 97-pound weakling, but now we prefer the magnum in the glove box.

After practicing with the .38 and then the Glock, trying futilely to match Pete's clean, tight, lung-shaped grouping on a new target, I'd had enough. My first and last experience with a handgun had been a breathtakingly fast love affair; once consummated, suddenly over. I'm glad I shot. I don't think I'll shoot again. I've hung up those machines for good.

An afterword:

When these words were first published, in *Popular Science* magazine, they resulted in enough hate mail from gun nuts that I had to change my e-mail address, which had been listed at the end of the piece. I got tired of opening my e-mail morning after morning to learn that I was a faggot, a Commie, a homo, an asshole, a man unable to satisfy his wife, a pinko liberal, a Jew bastard (Wilkinson???), a sissy, and someone obviously not entitled to citizenship in the U S of A.

A few also called me a liar, saying the business of hammering a .22 cartridge and having the slug explode out was impossible if the shell wasn't in the firing chamber of a gun. But that's how I remember it, and whatever happened, it sure pissed off my father. ✹

Vern Raburn's Connie

✱

I HAVE JUST RETURNED FROM FLYING THE world's biggest private plane. No arguments, now: Forget about the sheikhs' Jacuzzied 747s and the gold-fauceted Boeings of industry barons, for I'm not talking about professionally flown corporate aircraft. Consortium-owned bombers like the Confederate Air Force's B-29—or, for that matter, Kermit Weeks's privately owned but non-flying B-29—don't count. And you can't claim that the several B-17s and B-24s in private hands should be considered, for Vern Raburn's Lockheed 749A Constellation has got them beat on wingspan, weight, and horsepower. To say nothing of sheer sex, exoticism, and mystique.

Those who have flown to the rumbling beat of 72 pistons the size of garbage pails driving flailing propellers, back when air travel was an adventure for reasons that had nothing to do with Semtex, can't help but remember "the Connie." Legend has it she—Queen of the Skies (always a she)—was designed by Howard Hughes. Actually, the wacko in the sneakers simply set forth TWA's speed, range, and passenger-capacity specifications for what it needed in a new transcontinental airliner and let irascible Kelly Johnson, creator of Lockheed's P-38 Lightning, design the bird. (In fact, the basic Constellation has P-38 wings scaled up roughly two and a half times, though the later Super Gs and other stretched models got a different wing.)

Legend also has it that the Connie's distinctive fuselage, a gleaming curve as erotic as Marlene Dietrich's outthrust leg, was airfoil-shaped "to provide extra lift." Also nonsense: the unique barracuda profile got the tail up out of the hammering prop wash at one end of the airplane and put the stalky nosegear barely close enough to the ground at the other so that the huge props had adequate clearance. Of course, we'll never see its like again, for it's far more efficient to build beer cans with wings than to let the wind sculpt a shape.

One thing the legends did get right is that the Constellation's classic tripletail allowed the airplane to fit into TWA's existing hangars. It also allowed several generations of Connie pilots to boast, "It takes a helluva man to handle three pieces of tail at once."

I'm one of them now, for Vern let me fly his airplane.

"Half the fun of owning this thing," Raburn claims, "is watching other pilots fly it. I'm a really lucky guy, financially, and I've always enjoyed the opportunity to let other people experience some of what my good fortune has brought to me." Raburn's good fortune began when he got a case of entrepreneuritis and quit a good job with the huge American manufacturing company, 3M. In 1976, he opened in Los Angeles one of the very first computer stores in the United States. "The idea that somebody could actually own their own computer was so amazing to me," he laughs. "About like somebody owning their own airliner, in fact."

Raburn went on to pioneer the concept of selling software individually rather than making buyers take whatever was bundled with the machine. As a result, a new company called Microsoft hired him as its eighteenth employee, to start its new retailing division. There he met Microsoft founder Bill Gates's partner, Paul Allen, and became president and CEO of the Paul Allen Group—owners of two major professional sports teams, largest outside shareholders in Steven Spielberg's DreamWorks studio, provider of an ESPN content-based Internet site, and investors in everything from software design to modem and wireless technology to near-zero-emissions vehicle research.

Raburn's Connie, a "short" Model 749, never served as an airliner. She was delivered to the U.S. Air Force on the last day of 1948 as a C-121A, and immediately went to work shuttling across the Atlantic from Massachusetts to Frankfurt, carrying supplies that then went to Berlin as part of the famous Airlift. She was then fitted with a relatively plush (for the military) interior and used as a VIP transport for two decades before being sent to the boneyard in 1968—a time when riding in a Connie must have been regarded as punishment rather than reward for any "VIP."

In 1976, s/n 2601 came out of the boneyard to be refitted as a spray-plane, in which guise she served for eight years in Canada, dousing pine forests with spruce budworm poison. Her next moment in the sun came in 1984, when actor John Travolta bought the run-out hulk. Travolta was a boy when the Connie still ruled the skies in the mid- and late 1950s, but he vividly remembers them muttering westward out of New York's LaGuardia Airport, clawing for altitude as they overflew his family's New Jersey home. Newly prosperous, Travolta was *Saturday Night Fever*-ish to restore the airplane, but that remained for Vern Raburn to do.

"To me, the interesting triad is technology, commerce, and society," Raburn muses, "and the Connie is one of those fascinating loci where all three come together. This was the airplane that the airlines really started making money with. This is the aircraft that said, 'Here's what air travel is all about.' The DC-3 was the airplane that proved you could fly with speed and safety, but the Connie took it the next step: speed and safety plus comfort.

"It wasn't the jets that killed the ocean liners, it was the Connie. The Connie was the first airplane that could get above most of the weather, that eliminated those days when half the passengers were barfing, the days when the stewardesses passed out gum to chew so your ears wouldn't block up, the first one in which you could cross the country without stopping to refuel somewhere like Tulsa."

"She looked like an anteater on the ground, but she was so pretty in the air," recalls Helen Emberson, of Connecticut, a former TWA hostess who worked aboard transcontinental and transatlantic Connies in the 1950s. "Oh, it was the golden age. It was slower, so you had more time. The passengers weren't as rushed. You had a chance to talk to people. It was more genteel, and the appointments were just as comfortable as they could make them. For a long time, our Connies were all first-class. The first time I saw a coach configuration,

I thought, my gosh, with so many seats, it looks like a theater. But it was nothing compared with what it is now.

"People dressed up to fly. At least, they dressed as though they weren't going to the health club. I remember how horrified I was the first time I saw somebody come on wearing jeans. Noisy? Oh, no. If you sat right up over the wing it might be a little noisier, and there were a lot of deaf pilots, but it wasn't that bad."

TWA awarded its first-class Connie passengers colorful, hand-lettered "Transatlantic Flight" certificates, with space at the bottom for up to eight gold stars denoting repeat trips. Air France made its Connie ocean flights even more special. Each pair of the 18 seats reclined and flattened to become a double bed, giving new meaning to the luck of the draw in determining a seatmate.

Katie Morvay flew as a TWA hostess in the mid-'60s, when the Connies were aging stars upstaged by TWA's brand-new 707s and Convair 880s, but Morvay, who today lives in Maryland, still preferred the beat of the big pistons. "There was an impersonal feel about the jets," she recalls, "because you had two or three times as many people on board, and you couldn't give the kind of service you could on a 47-seat Connie. Yeah, they were one-class airplanes—all coach—but up at the very front, ahead of the galley, there was some sideways seating. The people who sat up there tried to sit in them all the time. I had one group of six businessmen who rode from Columbus to Washington in them every Friday afternoon.

"We served full meals on most legs, and you got to know some of the people. It was very special to fly back then, even for the short runs. You tried hard to make it really interesting for the children, because you wanted them to have an especially good experience."

But it wasn't all graciousness. Constellations, with their primitive pressurization, often flew at lightplane altitudes on shorter legs—10,000 to 18,000 feet, smack in the turbulent clouds. "I remember one lady who threw up eight times between Dayton and Columbus," Morvay muses, "and the circulation in those planes wasn't real good. So if somebody was sick, it permeated the cabin. We carried an awful lot of ammonia ampules."

Denver aerospace engineer Ralph Jones, a college student in Atlanta in the late 1950s, traveled home to Miami during school breaks and always chose TWA Connies. "First-class was in the rear, where it was quietest, so I'd sit all the way forward. The nose taper gave a small amount of forward view, and I could see the glow of Miami on the horizon a half-hour out.

"I picked night flights so I could see the exhaust flames. On takeoff, they would bathe the nacelles in great, orange-red plumes that snapped from side to side like a flag in a gale. Then they'd go to a short blue flame on the first power reduction, and finally, you could see them thin out when the flight engineer was refining the mixture one engine at a time.

"And the starting ritual was a show in itself. The first puffs of oily smoke would curl up in a confused ball, wondering where to go, until that monster prop whisked it out of existence. On takeoff, the noise was overwhelming, rich in specific sounds. Exhaust noise, prop vortices, clanks, pops, and whines as the landing gear came up one leg at a time, all manner of throbbing frequencies. I guess you could say the character of an airplane depends heavily on its parts count."

In the case of a Connie, it also depends heavily on its flight engineer. "Hell, anybody knows the flight engineer ran the Connie," laughs Martin Hall, an Alaskan who spent 8,000 hours doing exactly that. "We just needed those other guys to put the gear up and down and answer the phone. Oh, they did steer the plane a little bit, but it never seemed to tire them out."

"It is a flight engineer's airplane," Vern Raburn indeed warns me when I arrive for our mission. Raburn has a professional crew—two pilots and an engineer, plus four permanent maintenance technicians—that ferries and maintains the airplane for him. "The pilots just point it. Touch the throttles and you're in trouble with the engineer." The flight engineer's position in most of today's jets, if it exists at all, is largely the bookkeeper's chair, from which a baby pilot can observe line operations without getting in the way (the Concorde and early 747s excepted). Back in the era of big recips, flight engineers were professionals. "If you ever walked up to one of them and said, 'Congratulations, now you can be a pilot,' he'd punch you in the nose," Raburn points out.

N494TW's F/E is Tim Coons, a slight, semi-toothless young man with a balding head of shoulder-length hair that gives him an unfortunate resemblance to the destructive comedian Gallagher. Coons, with long heavy-airplane maintenance experience on the North Slope of Alaska, is the kind of guy who can change a 350-pound Connie wheel and tire singlehanded on an empty taxiway in less time than it takes me to jack up my little two-seat Falco in a fully equipped hangar.

Coons leaves me in charge of the airplane, sitting nervously at his complex panel after he fires up the left inboard engine and scurries down the boarding ladder to check an oil leak that he hopes he's fixed. "Just pull the mixture if you hear me yell." All I can hear is #2 running at a loping 600-rpm idle, so slowly that it sounds like a pile of enormous tin buckets clattering round and round inside a clothes dryer, but eventually Coons returns.

"Yeah, it's running like that because I think we've got a bad mag at idle—see the differences in the CHTs there?—and we're only running on nine cylinders." Coons's workstation is a dark warren of gauges and levers, the power controls up inside a cubbyhole against the right wall of the aft cockpit. On the small linoleum desktop is the oscilliscope of an ignition analyzer. Its flickering, luminescent-green trace shows god knows what—the firing of the plugs?—and can be set to probe any engine, any cylinder, any point during the power cycle.

The L-749 Connie—the "Baby Constellation"—has 2,750-horsepower Wright R-3350 engines, but they aren't the infamous turbocompound Wrights of later Super Constellations. In an effort to squeeze even more power out of the already-complex engines, Wright eventually added three power-recovery turbines to each engine—essentially exhaust-driven turbines, like turbochargers, that fed their torque directly back to the crankshaft via quill shafts and gears. The earliest -3350s—particularly in Boeing B-29s—had incendiary proclivities, largely because they had hot-running cast rather than forged cylinder heads, carburetors that provided excessively lean mixtures to some cylinders, and a forward-mounted exhaust collector for the front bank of cylinders, which nicely superheated the cooling air as it entered the nacelle.

The turbocompounds were torches as well, usually due to oil leaking onto the literally white-hot turbines through the power-recovery shaft seals. ("When they first put the PRT engines on the Connie, they melted some of the upper wing skins," Raburn says. "They finally had to insulate them with metal jackets nearly an inch thick.") But the interim R-3350s, by that time fuel-injected and with a revised exhaust system, were apparently good engines—never as popular as Pratt's myriad R-2800s but hard to beat for torque; after all, each cylinder displaced just over three liters.

Preflighting, pulling remove-before-flight pins, attaching the nosegear-steering drag link, securing the doors, running pilot and engineer checklists, and starting the four engines takes a good 45 minutes. Imagining throttle/mixture/spark/prime lever-wigglings appropriate to the engines' complexity, I found to my surprise that the big Wrights start about as easily as my Falco's little Lycoming. It's all in the shouting, and the routine goes like this:

Says the captain, who can see the left engines and reach their overhead-console starter-motor switchs: "One is clear! One to turn! One turning! One, two, three, four, five . . ." When the count reaches 12 silent blades swinging past the cockpit window, the copilot, who can't see the left engines, is supposed to know enough to flip the magnetos on. There's an explosion or two, brief silence again, a sudden roiling gout of what seems to be coal smoke, some wheezing, more cylinders joining in, and finally, the tin-cans-in-a-dryer loping idle. "One running!" the captain calls out for the engineer, who is assumedly also waiting for cries of "One a mass of flames!" or "One spitting out a whole lot of oily parts!"

For the right-side engines, the copilot assumes shouter and starter-motor duties, and the captain is consigned to the mags. It's all quite glorious—far more exciting than the *tik-tik-tik-hummmmmmmmmm* and lazy erecting of instruments that signifies a jet's awakening.

It's another 15 minutes' worth of warm-up, checklists, and run-up until we turn onto the runway, and the routine isn't done yet. The Connie has "walking struts" on the main gear, to soften landings, and they must be positioned for takeoff. Follow me through, here: Descending to a landing, a Connie's main-gear legs thrust appreciably forward, pulled into that position by 10,000-pound-tension steel springs. At touchdown, those springs—plus the normal strut oleos, of course—help absorb the initial impact and wheel spinup by letting the wheels move aft about a foot, straightening the gear legs until they're vertical. During the rollout, the airplane "falls off" the straight-up-and-down gear position and the main legs revert to a forward slant.

So before takeoff, the gear legs need to be straightened, the airplane like a sprinter going into the "set" position on the starting blocks. The pilot does this by tapping the brakes as the flight engineer brings the engine up to 1,700 rpm, and how gently it is done is a mark of an experienced Connie captain. (How nicely he's able to cushion the opposite motion of the gear legs during rollout is another).

Ready to go now, you intercom the flight engineer for max power—never "takeoff power," which to a professional engineer means to take *off* power. The PNF calls out "airspeed alive" and then 60 knots, which is your signal to let go of the nosegear-steering valve with your left hand and go to rudder control. Your mate calls V1, which comes up at about 100, depending on weight (we were at 78,500 pounds, with 2,100 gallons of the total 5,820-gallon capacity aboard and about a dozen people). Start to rotate—it doesn't take an enormous pull—and you should be flying when he calls V2 (best engine-out climb speed.)

Positive rate, go for 120 knots as the gear comes up, which I'm told can be an odd-looking procedure as seen from below. Whether it's the result of misrigging or a heavily challenged hydraulic system, some Connies apparently bring their gear up one leg at a time, in fits and starts. Flaps up when the pumps are done with the gear, call Scottie in the engine room, and whistle up climb power, pick up 155 knots for the climb, and you're on your way to Paris, Cairo, Bombay . . .

Well, Rochester, New York, in my case—a 30-mile cross-country from the small airport where Raburn's Connie had just appeared in an air show. En route, we climbed to 6,500 feet and did some airwork and stalls. The Constellation's forward cockpit is no bigger than a midsize business jet's, for it is at the very point of the arrow. Yet the deck space behind the engineer's seat was big enough to hold three or four gawkers standing around watching while I flew. (The area once held a navigator and a rack of radios the size of a small refrigerator. Both have been dispensed with in Raburn's airplane.)

As in a DC-3, your nose is only a foot or so from the windshield and the nosecone can't be seen, so the view is actually excellent. It's impossible to see the engines, but there's a small porthole with a lens on each side of the cockpit that provides an "objects in mirror are closer than they appear" view of the nacelles.

Raburn and his crew cruise N494TW gently, at about 40 percent power, burning about 85 gph per engine for a 200-knot cruise. Back when gasoline was the cheapest part of running an airline, operators would cruise their 749 Connies at 65 percent power, earning another 40 knots at the expense of vastly increased fuel flow.

Clean or dirty, the Connie stalls gently enough that passengers were standing around chatting in the cabin—largely a cargo area with a dozen seats forward—while I put them through half-G parabolas on the recoveries at about 75 knots gear and flaps down, 92 knots clean. "We could deep-stall it," Raburn's instructor-pilot Bill said, "but not at this altitude. Besides, it hammers too much." That's okay, Bill, straight-and-level pays better anyway.

Bill Dodds has been flying Connies since the early 1950s and has 9,000 hours in them. "Get him to tell you about the time he was 26 years old, coming back from Korea with a Connie full of wounded, and he lost three engines 250 miles short of San Francisco," Raburn will later say. "He flew the last hour and a half in ground effect, went under the Golden Gate Bridge, and was barely able to turn right and land."

Okay, let's see how much of a man it takes to handle three pieces of tail with all four running, with Rochester's 8,000-foot Runway 22 as a playground. The controls are well harmonized—heavy, heavy, and heavy—yet they're no heavier than they have every right to be. The airplane is . . . let's call it "ponderously responsive." Since it's a big P-38 wing, the ailerons have the hydraulic boost that was originally developed for the Lightning as an anti-flutter mechanism.

Downwind at 130 knots with takeoff flaps and call for gear down, turn base at 130 and slow to 120, go to 110 on short final and 1.3 Vso at 50 feet, then begin looking for the ground. As the airplane slows, it becomes increasingly important to lead corrections with firm rudder and crank in lots of aileron, but roll it back out as soon as the airplane begins to respond, else you'll get into a wallowing series of overcorrections. But don't be shy with control inputs.

At 500 feet, the F/E reaches up and snaps on the emergency hydraulic pumps, which will at least give you rudder and elevator—but manual reversion on the ailerons—if anything goes awry.

Bill is calling for the power settings, since I'd have no idea what to ask. "Give me 25 inches, Tim . . . 20 . . . 28, 28 [the bozo's too slow] . . . 100 bmep [on the F/E's torquemeters] . . . 20 . . . ease it off . . ." Some pilots, including Raburn, take over the throttles themselves at 500 feet, but Dodds's technique is to leave it all to the engineer, who can make or break a landing. "We could never see much out of any of the windows," ex-F/E Marty Hall recalls, "even that little lens-window. You could see more in the reflection in the fuel-flow gauge faces. If you were turning base, you'd see the runway through the captain's window reflected in the glass gauge. Elsewise, all you saw was sky.

"As for how we knew when to roll off the power, you could feel her float and settle. The sound changed as she got close, and you'd just know. Then again, you could make her drop out of the air by opening the oil-cooler doors and cowl flaps too soon. Or just as the wheels were about to touch, suck them all back in. The drag just disappeared, and you'd need another 5,000 feet. We could almost land her ourselves from back there."

My first landing is adequate, though on the go I have my feet a bit too high on the rudder pedals and get to touching the brakes, establishing a weave that must have any passengers far aft whacking from wall to wall. The second touchdown is a perfect squeaker, earning me a congratulatory handshake from a fellow Connie first-timer, TV producer Ed Shipley, who usually flies his F4U Corsair, P-51, and T-6—his very own three pieces of tail.

Between my stints at the yoke, I wander the ancient cabin. Peering out at the seamed and jointed old nacelles that are thrumming us along, I can see what ex-passenger Ralph Jones meant about character depending on parts count. We've become accustomed to glancing out the window at the polished silver potatoes that encase a jet's engines, yet there is no drama, no movement, no sense of thrust. Press your nose against a Lockheed Constellation's porthole, however, and you're face-to-face with primordial power.

Black oil creeps and spiderwebs back along the battered aluminum, thankful to be free of the inferno inside, and a variety of flaps and louvers vibrate in the raging torrent of air thrown back by the massive, geared, slow-turning propellers. The nacelles are dented by nearly half a century of flung ice, runway rocks, and mechanics' mallets, and exhaust pipes the size of household sewers are so hot they're white-orange at night. It's impossible not to sense the barely restrained mechanical hysteria being created a few feet from your face.

"The DC-3 you flew because you couldn't bear the thought of spending 10 hours on a bus," Vern Raburn says. "The Connie you flew because you wanted to. This airplane was the ultimate refinement of the piston airliner." And a reminder of the adventure, thrill, and specialness of travel that has been swept away by efficiency and ubiquity.

An afterword:

Vern Raburn is today fully involved with airplanes, not just as a vintage-airliner owner but as the president and CEO of a company called Eclipse Aviation. In late 2005, Eclipse is about to begin manufacturing a small, inexpensive (by bizjet standards) six-seat jet, the Eclipse 500, that Raburn sees as the core of new national air-taxi system. Intended not for Gulfstream-ing plutocrats but for business-class frequent flyers, the plan is for city-to-city hops aboard a network of thousands of Eclipse 500s, seats to cost roughly what an airline first-class ticket costs, but to serve nonstop 10,000 airports that don't even have airline service.

Don't bet against raburn's Everyman's jet. Remember that when he started selling computers, only the Pentagon and big universities could afford them. ✿

Go, Karts!

SO I SPEND MY LIFE PLAYING WITH fast cars, and the first time I'm on a track with my 23-year-old daughter—the blonde backpacker with the Ivy League sociology degree—she blows me into the bulrushes. Almost literally: The downhill chicane leading onto the main straight at Oakland Valley threatens to launch you right over the rumble strip into a cattail-bordered pond if you don't get the kart rotated and the power down early.

That's right, go-karts. Those things that Americans think of as Cushman-engine amusement-park rides, but that Europeans and South Americans employ to turn teenagers into the world's finest open-wheel racecar drivers. (Every Formula 1 driver of any consequence has been a national or international kart champion.) Brook and I had entered a four-hour kart endurance race at Oakland Valley Race Park, near Port Jervis, New York, in that bosky corner of the state where Pennsylvania, New Jersey, and New York meet. I came home wishing I'd discovered karts 50 years ago. Never have I learned as much about car control in as short a time.

Karts look ridiculous, with their wheelbarrow tires, primitive frames, lawnmower engines, and sitting-in-a-hole seating position, but don't be fooled. The tires are miniature but real Bridgestone racing slicks, the frames work just fine, and in the faster classes, the engines accept the same degrees of supertuning that a motorcycle does. (Okay, the seating position does look ridiculous, although in the fastest "laydown" classes, the driver is as recumbant as a luger.) The combination somehow replicates all of the dynamics of a true mid-engine racecar except the sheer acceleration and speed, though the fastest laydown shifter karts can do 140.

You're going to laugh, but the racecar that I'm rhapsodizing about was powered by a nine-horsepower, four-stroke, box-stock Honda industrial engine, had a single hydraulic disc brake on the rear axle, and topped out at all of 40 mph. But it's like going 80 in a small speedboat. Unless you've been there, you can't understand how different the sensation of speed is when your buns are two inches off the asphalt and the track is tight enough that only near the end of the main straight does the kart top out.

Karts respond almost exactly as a pure racecar does to threshold and trail braking, to clumsy steering inputs, and asking the tires to do too much, to left-foot braking, to throttle steer, and maintaining the car's balance. One of the things they'll teach most effectively is that a vehicle balanced, pointed, and under firm power is a lot more controllable and predictable than is one being driven cautiously, tentatively, and fearfully. And it's a hell of a lot faster.

Fortunately, my daughter and I were teamed together on the same kart, along with a third driver from our local Porsche Club chapter, or I'd have been really embarrassed to try and match her times. But we did drive "against" each other in separate karts during a morning-long training session run by an acerbic Belgian, Jay de Marcken, who travels all over the East running races with a fleet of spec-built beginner karts and giving introduction-to-karting classes. De Marcken lectured from atop a rusty mountain bike, the tires of which hadn't seen an air pump in years, and we walked the 11-turn, half-mile track while he pedaled the arc from turn-in to apex to track-out, trying to keep the attention of yet another bunch of karting novices, some of whom listened while others gossiped and cracked wise. How hard can this be, they thought? Little did they know. (Go to de Marcken's site, www.start racing.com or the track's, www.ovrp.com, to learn all you need to know to try it yourself at some of the best karting facilities in the country.)

When we ran the actual race, the field off in a tight rolling grid late that Saturday afternoon, Brook did the crowded start, drove the lion's share of stints—there were 11 mandatory driver changes and two refuelings—and took the checker. She notched our team's fastest laps by far, ran with youthful consistency while Patrick and I skittered and spun, and kept faithfully unlapping us, retrieving the places that we guys lost every time we planted our middle-aged asses in the kart's stiff, unforgiving seat.

Talk about stiff—also bruised, cramped, and brain-dead—which was what all three of us were after four hours of racing. It gripes me when jock-sniffing big-city sportswriters, all of them professional spectators and most of them wannabe stick-and-ball-game players, announce that racecar drivers "aren't athletes." Strap on a kart for a total of an hour and 20 minutes and you'll get a hint of what the stresses must be like in a real racecar. Indeed, Pat wilted after the third hour and admitted that he was too tired to keep the thing on the track, and my blonde took up the slack. "Oh, this isn't really an endurance race," de Marcken's assistant later laughed. "Most of our long races are eight, 12, or even 24 hours. We did a 24-hour down in Florida a while ago, and a 10-year-old and his brother won it." There's something to be said for a good power-to-weight ratio.

One nice thing about karts is that they have no rearview mirrors, so I was spared the intimidation I normally feel when, as a rank amateur, I see a car close behind me on a racetrack. The faster kart drivers—and there were many—did announce their desire to pass by belting my rear bumper hard enough to make me see stars and curse, but unless you glimpse a front wheel in your peripheral vision, just work on your line, braking, and power, which is as it should be. At least, that's what I thought until some bozo straight-lined a chicane and in a split second launched himself over my front end and out into the dimly floodlit off-track darkness as the final half-hour of the race unwound.

I was embarrassed to see Brook's "team fastest lap" notation every time I checked the big video monitor that tracked the race positions in the Oakland Valley Race Park clubhouse—and it kept decreasing by hundredths and tenths—yet at the same time I couldn't have been more proud as I watched her wail past in her tiny racecar, lap after lap. "You should have seen the look on her face when the guy chasing her at the end of the race came by our pit and told her she was fast," Patrick later said.

In a country where size matters, low-horsepower karts have never gotten the respect they deserve as competition machinery. Which is a shame, since it's the cheapest, simplest, safest, and most effective way on the planet to go racing. ✿

FORMULA NONE

✵

FORMULA 1 CARS ARE THE MOST COMPLEX, most electronically sophisticated, and most technologically advanced racecars on the planet.

They are also the most boring.

Damon Runyan once wrote that sailboat racing was as exciting as
watching grass grow. Having recently attended the Grand Prix of San
Marino, in Italy, I have decided that F1 is as exciting as watching time-lapse
photography of grass growing: considerably faster, momentarily fascinating,
but ultimately tedious.

The momentary fascination is the sound of V10 piston engines turning
turbine speeds—yes, almost 19,000 rpm in some cases—at which the titanium
connecting rods actually stretch tangibly enough that the slight elasticity has
to be taken into account lest the demitasse-size pistons smite the heads. That's
an explosion of fuel in each combustion chamber 150 times every second.
When you have listened just once to a full field of such cars winding up for
the green light, you're probably two years closer to needing a hearing aid.
What you hear on the Speed Channel's TV broadcasts of F1 races is a wimpy
whisper of actuality.

In the time that it takes a good sports car to accelerate from zero to 60,
an F1 car will be doing 150, even though it'll be geared for 220-mph road-
course straightaway speeds. The McLaren Mercedes team used to show fac-
tory-tour visitors a film in which a race-prepared A-Class Mercedes laps
Silverstone. As it approaches the final turn before the finish line, driver David

Coulthard puts down his cup of tea and a copy of the *Financial Times*, straps in, and sets off in pursuit, circling the entire track and passing the Mercedes just before the little subcompact completes its lap.

Yet vastly more powerful than their 2G acceleration is an F1 car's braking force, which can generate a near-incomprehensible 6Gs. If a 170-pound driver's six-point harness were for some reason to fail under braking, he'd be slammed against the steering wheel with a force of over half a ton.

The steering wheel of Michael Schumacher's Ferrari is the diameter of an ordinary dinner plate. Encompassed within that small plot of real estate are more than two dozen buttons, knobs, selector switches, and rheostats with which Schuey can vary most of the parameters of his racecar's operation. Plus a small computer screen onto which a wide variety of data are displayed for his amusement. I have flown airplanes with fewer controls.

There was a time not long ago when Formula 1 organizers feared that their racecars were becoming so highly computerized and automated that the inevitable next step would be transferring absolute control from the driver to the engineers and computer geeks in their pits, who could easily radio-control the throttle, brakes, gearbox, and steering more rapidly, efficiently, and safely than could he. No, they wouldn't remove the driver, since there had to be a sacrificial lamb in the cockpit to make things interesting, but he'd basically be like the kitten a kid might jam into his r/c model car.

After all, transmissions had reached the point where accelerating from a standing race start was simply a matter of pushing the "launch" button. Wheelspin accelerating out of corners was eliminated by very sophisticated traction control. And antilock brakes meant a driver simply needed to brake as hard as possible without fear of lockup. Stand on the gas, stand on the brake . . . all that needed to be added to the equation was automatic steering.

So some of the electronics—mainly traction control and two-way telemetry between car and pits—were outlawed, though suspicion remained that at least for a while, some teams retained the necessary wheelspin software deep inside their engine-control modules. (Sharp observers could detect traction control, they felt, by the way the engine sounded coming out of corners.)

The reason that Nascar decrees that its racecars use simple, old-fashioned carburetors rather than far cleaner and more efficient fuel injection is not that Nascar loves carbs, but that modern fuel injection requires a computer. Nascar knows that once a microprocessor is in place, cheating can no longer be discovered with micrometers, templates, experienced eyeballs, and knowledgeable hands, but will require people with the time and talent to laboriously decipher computer code.

The reason that Ferrari wins so consistently in Formula 1—six world championships in six years, by the end of the 2004 season—is that they have the best computers. Yes, Schumacher is the best racecar driver in the world, and that helps, but the reason he's best is partly raw skill and partly his innate, cyborg-like computer awareness. Decades ago, a talented racecar driver was a guy who had "a feel for the machinery"—someone who could read the messages sent by a car's engine, transmission, and suspension. Today, he needs that plus a virtual USB port in his brain.

There was a time not long ago when I would, despite my reservations about the sport, religiously watch F1 races on television. Even to the extent of getting up at one and two in the morning to watch a live race from Australia or Japan. No sport has as thoroughly and intimately put itself on camera, since the bulk of its revenue and sponsorship support comes not from the tens of thousands of fans in the stands but from the tens of millions in front of TV sets all over the world.

In Europe, if you buy a subscription to the special satellite channel, you can watch BernieVision, so-called after F1 rights-holder and billionaire Bernie Ecclestone. Two 747 freighters are required to fly to each race the necessary equipment to broadcast this program, and there are cameras everywhere. They show the view from the driver's nostrils, from the pit crew's crotches, from every corner, even, if you choose to not even bother watching anything as coarse a racecar, simply a view of the computer readouts that trace the race.

But I've stopped watching. I realized that since the end of each race was preordained—Schuey first, Rubens Barrichello second, everybody else also-rans—I'd been turning on the TV simply to see the almost inevitable first-lap clusterfuck, the multicar crash that would wipe out 20 percent of the field before they'd raced more than 20 seconds, and that suddenly struck me as terribly wrong.

The tens of thousands of screaming, flag-waving, airhorn-farting Europeans at the Imola track the day I was there might have wished for more action—there were just two passes of consequence during the entire race that I saw, one of them in the first few hundred yards of the first lap—but for them, the outcome of what was essentially a high-speed parade was still satisfying. The two hugely powerful home-team Ferraris were one-two in the same order in which they'd started, the Williams BMWs on slightly less effective tires were three-four ditto, and everybody else was a supporting player.

It's a culture thing. Americans demand more drama. We want a 45-to-42 seesaw Super Bowl; Euros are happy with a 1-zip World Cup game after an hour and a half of footie. F1 fans feel that the best car on the track should win the pole position, should therefore be first through the first turn, never be passed, and finish first.

The drivers who pilot these three-liter, 870-hp, 1,150-pound rolling missiles are the elite, but they are a sullen lot, giving away autographs as freely as they do thousand-Euro bills. Most make Russell Crowe look as giddy as Pee-wee Herman. For Americans, drivers come first. "Ricky Rudd's my man," a Winston Cup fan will say. Or Mark, or Rusty, or Jeff. Whether they're driving a Dodge, Ford, Chevy, or Pontiac is important, but we like to think that American racing is athlete versus athlete, not a battle of countries and corporations. Before each Winston Cup race, the drivers are paraded around the track each in his own signboarded convertible, each alongside his own local beauty queen. At an F1 race, the drivers are crammed like cattle onto a single flatbed. Are they trying to tell us something?

For Europeans, the Constructor's Championship is arguably as important as the Driver's Championship. So the Italians got pushed around by the Germans in World War II? Stick this up your nose, Mercedes, say the Ferrari fans. So you think your Spitfires were so slick, say the Germans? Lead, follow, or get out of the way, puny Jaguar.

Which is amusing, considering that nearly every F1 car except the Ferraris is made in England. The top teams develop their cars in full-scale, rolling-road wind tunnels, and the downforce their aerodynamics generate is so enormous—about 2,600 pounds at 150 mph—that the cars could easily race upside down on an inverted track, as long as it didn't have any corners so slow the cars fell off the ceiling. (If F1 were an American sport, some promoter would be sure to try it.)

Formula 1 is the most exclusionary sport in the world. Polo seems positively y'all-come in comparison, for Grand Prix racing reflects Europe's near-universal class condescension. The commoners sit on bleachers, and the pits and paddock are for the lords and ladies. My temporary Paddock Club "membership" at Imola cost my hosts—Mercedes-Benz—nearly $4,000, they revealed, but even with four large, you couldn't buy your way in. Unlike Winston Cup's buy-'em-at-the-gate paddock passes, you'd have to be invited by a team or sponsor.

At an F1 race, celebrity-sniffers congregate around the impenetrable, electronically controlled Paddock Club turnstiles hoping for a glimpse of a driver returning from a brief, painful PR appearance in a sponsor's hospitality suite, for that's the only way they'll ever see one up close. As I came out of the Club on the afternoon of qualifying, a group of Italians pointed at me, clapped hands to mouths, and jabbered excitedly. Somewhere, apparently, there is a white-haired *ancien* from the days of skinny-tired, front-engine Grand Prix cars who must also wear Harry Potter glasses and a goofy-looking Tilley sunhat. Knowing F1's rules, however, they didn't ask me for autographs.

FLYING FORTRESS

❖

THE MOST MOVING SCENE EVER CAUGHT BY Hollywood's cameras plays near the beginning of *Twelve O'Clock High*. (Don't argue. I get goose bumps every time it unreels in my mind.) Dean Jagger, playing an American businessman-tourist in late-'40s England, stands on the weedy, cracked concrete of what was once his unit's bomber base. The air is windless, a day of sunny summer postwar prosperity. Until first one, then another, and another screech breaks the stillness. It is the eerie wail of aircraft inertia-starter clutches cranking big Wright R-1820 Cyclones, a squadron's worth awakening like phlegmy old men hacking into their handkerchiefs, the blast of their propeller blades now flattening the grass at Jagger's feet.

It is 1943 again, and the airplanes idolized by those of us in America who were only kids during the war are off to macerate Festung Europa yet again. Broad-shouldered, high-tailed Boeing B-17 Flying Fortresses are awake, angry and anxious to bomb Berlin back to the Stone Age.

Forty-nine years after first seeing that film with my dad in a suburban New York theater, back in the days when a moviehouse played just one movie, I'm climbing into the captain's chair of the exquisitely restored B-17G *Yankee Lady*, about to clutch a semicircular control yoke big as a garbage-pail lid. I am fulfilling every middle-aged American pilot's dream. Scott Smith, who flew 17 missions over Germany in Forts, knows the feeling. "When I was in the left seat of a Seventeen the first time, much as I hated the idea of four engines—we all wanted to be fighter pilots—I remember thinking, 'Man, I've got 4,800 horsepower in my right hand. I could beat Joe Louis.' "

It is a hot spring day in peaceful Ypsilanti, Michigan, where the group of largely geriatric volunteers who call themselves the Yankee Air Force—and Scotty is one of them—runs a gallant little aviation museum at the Willow Run Airport and polishes its small fleet of flying warbirds. On the ramp beside *Yankee Lady* is a glittering, polished-aluminum North American B-25 Mitchell medium bomber and a doughty Douglas C-47. Across the field, appropriately, is the vast factory where the Ford Motor Company once built thousands of B-24 Liberators.

The Yankees are reluctant to let me pilot their prize. It is the best-restored of the 13 Flying Fortresses still capable of flight—all that remain of the 12,731 originally manufactured—but the club has already had an expensive exposure to inattentive aviators.

Several years ago, a retired U.S. Air Force captain was granted the left seat of their C-47. He lost control during a takeoff in front of the EAA convention crowd at Oshkosh, groundlooping the poor Doug so enthusiastically that the landing gear collapsed, the propellers smote the ground, and both engines were ruined. (I was there, and until he hit the broad ditch that did the damage, it looked as though he'd be taking out a fair part of the flightline crowd, too.) It cost the Yankee Air Force over $100,000, and as far as they're concerned, tricycle-gear pilots like me are anathema.

So I will fly, but I won't land. Okay by me, since *Yankee Lady* is uninsured but valued at something north of $2 million. Right now, the Israeli government is setting B-17 values. Israel operated three surplus Flying Fortresses during both its 1948 and '57 wars, and even bombed King Farouk's palace as part of their clandestine delivery flight from Czechoslovakia, but the Boeings were scrapped. The Israelis apparently are looking for a restored example for a museum. Israel offered $1.5 million for *Sally B*, a less thoroughly restored example, and rumor has it, it might go as high as $2 million. Which, the Yankee Air Force feels, might make their bird worth . . . oh, how does $2.5 million sound?

From the outside, a B-17 looks enormous, at least by 1940s standards. Its tires are the size of an earthmover's, the cockpit with its Gar Wood-ish speedboat windshield perches far above, atop that glass-tipped bombardier's nose pointed proudly skyward even at rest. (Forgive me, but B-17s inevitably color prose purple. Consider this, from a February 1943 *Life* magazine article on the famous Pacific-theater Fort, *Suzy-Q*: "There's just one thing she wants to do and that's to kill Japs. She knows her big job is to lay a string of bombs on an enemy ship or airdrome and knock equipment and men to hell in a thunderous boom. But she also likes to snuggle in low over a target and, with machine guns blazing, pick off every damn Jap in sight. Then she sticks her blunt nose up toward the sky once more and hightails for home hundreds of miles away.")

The only advantage to being a B-17 gunner was that you got to board through the aft door, just a long step up from the ground, rather than performing the chin-up and back somersault through the nose belly door into officer country. How the lieutenants managed such gymnastics, particularly in flying gear, is beyond me. But it wasn't just a Hollywood invention, for I remember how pleased my old boss Bob Parke was when he discovered that, like riding a bicycle, he could still do the flip a quarter-century after he'd last flown a Fort, when he got a chance to fly a restored example while he was the esteemed editor and publisher of *Flying* magazine.

But we're all too ancient for that—71-year-old Yankee Air Force left-seat pilot Don Harner, 76-year-old instructor pilot Dick Bodycombe, 66-year-old flight engineer/mechanic Norm Ellickson, and me, at 62, the kid.

The cabin door leads into the broadening of the tailcone. To the left is the dark crawlway to the tailgun, a contorted squeeze past what looks like a boulder. It's actually the lumpy canvas bag covering the tailwheel and its retraction mechanism. There's a separate, more direct exterior door for the tailgunner, but it wasn't often used for entry: nobody wanted to sit back there any longer than they had to. The gunner straddled a bicycle seat and kneeled on a pair of pads for hours at a time as it was, lonely as a monk at Mardi Gras. Almost worse than the bullets was the breeze. Every bit of superchilled, high-altitude air that poured into a B-17 through a variety of open windows chinks, gaps, leaks, and shrapnel holes ultimately found its way back to the tailgunner.

To the right, the waist gunners' compartment widens out, but not enough for two men to fight back to back. Amazing how long it took for Boeing to figure out that the gun windows should be staggered—until the final model, the B-17G, was introduced—rather than being symmetrically located and thus placing the waist gunners bum to bum. *Yankee Lady* is a G. In fact, she's the youngest Flying Fortress in the world—the thirteenth from the last one manufactured—so her waist windows are not only staggered but also glassed-in, which was a big improvement over earlier models. Being asked to fight during a European winter while standing at an open picture window six miles above the earth shows a certain failure of common sense on the part of the B-17's designers. Who, of course, had probably never been to 30,000 feet in their lives.

When you climb into a Fort's cabin, it becomes chillingly apparent that this was a mean, miserable, vulnerable, uncomfortable, horrifying place to go to war. I can't help but recall one B-17 crewman describing the shrapnel of nearby flak bursts as sounding like a constant rain of nails flung at a tin roof, for the airplane is naked inside—everything exposed and only sheets of thin aluminum cocooning the crew. You move in a half-crouch, yet it's impossible to avoid whacking your head on something too close above—bulkhead, gun mount, radio rack, ammo chute.

The radio operator's station is just aft of the bomb bay, and in earlier B-17 models, Sparks was given a single ineffectual .50-caliber gun to fire when he wasn't answering the phone. G models dispensed with the weight of that gun and its ammo, and in fact, *Yankee Lady*'s radio-compartment upper window is a predecessor of my Toyota's removable sunroof: pull it out, stow it away, and ride through the Michigan spring with our heads in the 150-mph breeze, peering back at a vertical fin that, from this perspective, is thicker than a Piper Aztec's wing.

The bomb bay takes up the fattest part of the fuselage, and passage through it to the cockpit requires tightroping along a narrow aluminum plank above the bomb-bay doors. A tunnel below the cockpit attains the navigator's and bombardier's stations, and it was here that hell looked you in the face. A cagy Luftwaffe squadron leader, Egon Mayer, discovered that if he attacked a B-17 from directly in front, he was below the fire arc of the top turret and above the belly gun's, opposed only by a terrified bombardier's single .50. And the bomb guy hadn't had much time to learn how to aim it. Mayer's maneuver, a German invention as innovative as the Thatch Weave, eventually led to wave after wave of 12-abreast Luftwaffe fighters going head-to-head with B-17 groups. Even though *Yankee Lady*'s nose is pincushioned with four .50-caliber machine guns—the G model's eventual reaction to this frontal-attack threat—Messerschmitt and Focke-Wulf drivers hunkered down behind bullet-proof windscreens and the huge metal lumps of their engines. B-17s might as well have had Tupperware for armor, since the entire nose is simply plastic.

It surprised me that riding in the glass nose perched on the bombardier's little swivel seat is not the otherworldly experience I'd expected. I suppose that seven years under a Falco's bubble canopy had accustomed me to almost as good a view, and the fact is that it's still a small, constricted space filled with weaponry and head-banging equipment. In the G model, the bombardier had to know a little more about gunnery: when he wasn't bomb-aiming, he swung into place a yoke-like arrangement that steered the twin .50s beneath him in the chin turret, while his navigator companion had the choice of calculating wind drift or alternately firing the two hand-operated cheek guns. (Chin, cheek, nose . . . would the H model have had ear and forehead guns?)

So the pilot's seat is the best in the house, especially when you've waited 49 years to be Gregory Peck. The brakes gripe and squeal as we taxi out, and the old bomber nods along in elephantine fashion, the expansion cracks in the taxiway setting up a ponderous porpoising. Takeoff checks complete, flight engineer/mechanic Norm Ellickson, the retired director of maintenance for Northwest Airlines, stands between the pilots' seats to set final power with the unusual thicket of throttles. They're the only ones I've ever seen that you use with your hands upside down, palms skyward. They're uniquely arranged so that an easy handspan can vary just the two outboard engines, for more precise formation-flying.

Yankee Lady, its crew boasts, is the only restored B-17 flying with a properly set-up quartet of turbochargers. (Some-17s have even dispensed with their turbo plumbing entirely.) Like other turbos of the era, its General Electric turbines are open-air, the exhaust blowing straight through the vanes and out the bottom of the unit rather than spinning the wheel inside a housing. What this loses in efficiency it gains in cooling of the vanes. It also allows for a blast tube that directs slipstream at the central bearing. (GE learned fast, though; such turbos were the beginning of the technology that led to its leadership in jet-engine building.)

The turbochargers on a G model are controlled electronically by a rotating knob atop the power pedestal, marked with numbers from zero to 10, corresponding roughly to added inches of boost. From eight to 10 is red-arced. "It's fine to use if it saves your ass," Dick Bodycombe says, "but it means four engine changes." The Yankee Air Force uses seven for takeoff, which means full throttle gives 42 inches and 2,500 rpm. (The Cyclones also have mechanical superchargers, so even with the turbo controller set to zero—wastegates wide open—the engines will put out 35 inches.)

Takeoff has a ponderous, are-we-flying-yet inevitability to it, and I wonder what it must have been like to lift one of these beasts off the ground with two tons of bombs and gas to go to Berlin and back. We're light, with phony aluminum guns, no bombs, a local-flight load of 100-octane, and a "crew" wearing T-shirts rather than steel flak helmets and electrically heated flying suits. Gear and flaps up, we go to METO power, and Bodycombe cranks the controller back one inch. Climb power calls for three inches of boost, and for cruise, the turbos are shut down entirely, as an economy measure.

Our mission, if we choose to accept it: buzz the Azalea Day parade on Grosse Ile, in the Detroit River, and then land and sell rides at $400 a seat to whoever bites. Actually, they're not called rides but "flight experiences." *Yankee Lady*'s "limited" certification doesn't allow carrying passengers for money, so the Yankee Air Force turns seat applicants into new YAF members and "gives" them the trip aloft as a membership privilege.

The YAF supports its little fleet by such modern-day barnstorming. *Yankee Lady* costs over $2,000 an hour to operate, so they travel the country every summer displaying the airplanes and selling everything from T-shirts and tours of the cabins to flight experiences. B-17 rides are particularly popular, for the Flying Fortress legend looms far larger than does the truth. "Fighters weren't too happy about having to attack one of these things," announces a spectator full of expertise after we land at the Grosse Ile Airport. "I'll tell ya, it was like whacking a beehive."

Well, not exactly. Much of the B-17's glory was generated by Boeing's and the Army Air Force's PR departments, and it was only as the end of World War II neared that the very latest models of the airplane achieved anything like their promise. "Flying Fortress" was a breathless sobriquet bestowed on the airplane by *Seattle Times* reporter Dick Williams, who attended the initial roll-out of an airplane that was anything but. Indeed, the RAF tried to operate early B-17s before the U.S. entered the war and quickly declared them useless for combat, relegating them to Coastal Command duties.

At Pearl Harbor, radar operators contributed to the carnage by assuming that the Japanese attackers on their primitive scope were a flight of B-17s due to arrive from California. When the lumbering Boeings did arrive, in the midst of the attack, they carried no ammunition. Japanese fighters turned most of them into scrap metal. Soon thereafter, our need for heroes created a more abiding part of the B-17's legend. To this day, warbugs will tell you that America's first World War II Medal of Honor winner was a young Army Air Force B-17 pilot, Captain Colin Kelly, "who dropped a bomb right down the stack of the Japanese battleship *Haruna*." Unfortunately, the *Haruna* didn't exist in 1942, and Kelly in fact bombed and damaged a cruiser but was quickly shot down by angry Zeros. He was awarded a posthumous Distinguished Service Cross for staying with his airplane in order to let his crew bail out.

In Europe, B-17s were forced to carry out a flawed aerial strategy: They would bomb with assumed impunity during the daytime, visually, in loose formations without a fighter escort, from a height and speed so great that no enemy fighters could reach them. (Remember Francis Gary Powers and his untouchable U-2? Some people never learn.) B-17s weren't able to reach the 35,000-foot-plus altitudes for which they'd been designed, and the Germans had supercharged fighters that had no trouble achieving the Forts' more common 31,000-foot ceiling. And 88mm flak found its way as high as 41,000 feet. "By the time I got there, the German gunners had five years of practice on live targets," recalls Scotty Smith, "and they were pretty damn good." Nor was it hard to find the B-17s, for the tracery of lovely exhaust-condensation contrails you see in the classic photos of B-17 raids pointed straight to them.

Okay, new strategy: tight formations, called combat boxes, so that every B-17's guns could cover every other ship in its element. One thing this unfortunately achieved was the horror of closely formatted B-17s dropping their bombs onto B-17s below them. Nor is it known how many B-17s riddled nearby B-17s in the heat of close-quarters combat. (A B-17's belly turret has solenoids that interrupt the firing mechanism if the gunner inadvertently wheels the gun toward his own airplane's propellers while tracking a fast-moving target. But there was nothing to keep him from hosing a wingman.)

At one point, the Air Force even tried to insert B-17 gunships onto the combat boxes, designated YB-40s and armed with a variety of extra guns but no bombs. The experiment was a failure. When the -17s dropped their bombs, they accelerated, and the YB-40s, unable to dispose of any of their load, couldn't keep up with them. The YB-40s did prove the efficacy of the chin turret and staggered waist-gun positions, however, and those mods were added to the B-17G. But the weight of the -G's extra guns, ammunition, and four dedicated gunners cut significantly into the airplane's useful load. The theoretical maximum bomb load of a -G was 13,600 pounds, but they rarely carried more than 4,000 over Europe, and usually were loaded with half that. It was a lot of manpower, engines, and effort to deliver just a ton of bombs.

Dick Bodycombe, my instructor in the right seat of *Yankee Lady*, flew 13 missions as a B-24 Liberator pilot, and I ask him how the American daylight bombers compared to the night-bombing Lancaster. "Did you ever look at a Lanc?" he laughs. "It's a flying bomb bay. The doors run the entire length of the fuselage. After the raids on London and Coventry, all the RAF wanted to do was kill Germans, not bomb factories. We had to have an identifiable, legitimate target before we dropped. The lead bombardier could get court-martialed if we didn't. The RAF, they bombed at night, one plane and one bombardier at a time, and they dropped on the fires set by whoever had preceded them."

A key element of American strategy was the legendary Norden bombsight, which has been called the single most complicated mechanical device ever manufactured. The Norden was a remarkably compact and precise analog calculator that compensated for everything that might affect the ballistics of a bomb—air density, wind drift, the bomber's airspeed, and groundspeed—and that was linked to the B-17's autopilot so that the bombardier could actually guide the airplane through the bombsight during the final run in to the target.

It was invariably said by enthusiastic AAF public-information officers that a Norden "could drop a bomb into a pickle barrel from 30,000 feet." Avers Don Sherman, a Detroit writer who has studied the Norden saga, "The Norden had only a 20-power telescope, so you couldn't even see a pickle barrel from 30,000 feet, much less hit it. You could make out a factory, but that was about it. It was also very easy to defeat the Norden when it was used at high altitudes. Smoke screens worked just fine, ground fog was a barrier, and the simple fact was that the year of the most disastrous B-17 raids—1943—saw an unusual amount of bad weather over Europe."

"I brought a lot of bombs back, or dropped them in the Adriatic," says Dick Bodycombe. "Killed a lot of fish, but that was better than trying to land with live bombs. The bombardiers were supposed to disarm the bombs if we had to bring them back, but some of them just didn't want to go into that bomb bay once the bombs were armed." Scott Smith admits that even when the bombs were toggled over a target, often it was literally a hit-or-miss proposition. "Some missions we went on, we wouldn't even have hit the ground if it hadn't been for gravity," he laughs.

B-17s didn't bomb Europe in any numbers until late January 1943. Within several months, things had gotten so bad that Curtis LeMay, later to become chief of our Strategic Air Command, announced before leading his B-17 squadron over Germany, "I'm going to have an easy job today. All I need to do is follow the trail of parachutes and burning B-17s." It would be a while before the Flying Fortress stopped being a Flying Fort Apache. By the summer of 1943, B-17 losses had become unacceptable, and the long-range big raids were ended.

They began again in January 1944, thanks to the development of fighters that had the range to escort the B-17s all the way to the target and back, and the firepower to swat the Luftwaffe out of the sky. By D-Day, in June, American fighters had pretty much swept the skies clean, and finally the Forts flew largely unopposed. Yet ultimately, it was the low-level fighter-bombers—mainly P-47 Thunderbolts—that destroyed Germany's transportation and communications networks and enabled the invasion of France, and then Germany. The carpet bombers had tried to level Germany's industry but had failed.

Germany even got one more chance to smoke the clumsy B-17 armadas. In May 1944, the Luftwaffe introduced the Messerschmitt Me-262, the world's first operational jet fighter. The Me-262 would have flown rings around escorting P-51s. Fortunately, Hitler—perhaps impressed by the devastation that American fighter-bombers were wreaking—declared that the -262 would itself serve as a bomber rather than as an interceptor. Game, set, match.

But all this is ancient history in the skies over Lake Erie in 1998. *Yankee Lady*'s four nine-cylinder Wrights mutter away lazily, the noise level in the cockpit surprisingly low, in part because each engine's exhaust runs through the turbo plumbing, which acts as a muffler. I peer out the left-hand cockpit window at an anachronistic sight that I've seen only once before, on an ancient Ford Trimotor: the B-17's basic oil pressure gauges are out in the breeze, on the inboard sides of each nacelle. "They were particularly useful during cold-weather operations," Bodycombe says. "You could trash an engine by the time the cockpit gauges confirmed a lack of oil pressure, what with all those long, tiny capillaries the oil had to feed through to get to the cockpit." Indeed, the first call after engine start-up is, "We have oil pressure," as the nacelle gauges snap to life.

On the pedestal is a simple power-setting table, obviously designed for people who had more important things on their mind than tweaking mixtures and prop pitches. Takeoff, climb, fast cruise, medium cruise, and long-range cruise each get a single power setting, but below them are two more notations: "Fly to target at 155 [mph] IAS 1,800/28"/160 gph. Return from target at 140 IAS 1,400/28"/120 gph."

B-17s were never particularly fast, but in their excellent book, *Claims to Fame: The B-17 Flying Fortress*, Steve Birdsall and Roger Freeman tell my favorite B-17 story, about a Fort named *Gremlin Gus II*, which had its entire topside cab and cockpit top framing removed to allow the loading from above of torpedos, the intended target being the Tirpitz. The project came to naught, and the top of the fuselage was replaced, though the cockpit framing and glass wasn't, leading to the only open-cockpit B-17 in existence.

"It is said that Major Ralph Hayes, one of the project officers, had a favorite prank," Birdsall and Freeman write. "Being stripped of armour, *Gremlin Gus II* was considerably lighter than a standard B-17 and, in consequence, much faster—an estimated 30 mph extra at top speed. It is alleged that Hayes would find a B-24 Liberator and position to the rear. Normally a B-17 was slower than a B-24, but not *Gremlin Gus II*. At an appropriate moment Hayes would advance all four throttles and shoot past the Liberator, and as he passed the cockpit of the B-24 he would stand up and salute."

I wouldn't have a spare hand to salute with, however, for the controls are heavy, and urging the airplane into a turn is a ponderous procedure: feed in way too much aileron, wait for the airplane to bank, then take a normal cut at the wheel. A B-17 has two independent sets of control cables to every flying surface, in case one is severed by enemy fire, and there is no hydraulic assist, no electric motors to do the pilot's bidding. "Remember, you're moving all that cable," Bodycombe tells me. "There's no boost. It's fun to watch our Yankee Air Force members who are senior airline captains get out of their 350-ton airliners and try to fly this 55-year-old technology. The first thing they do is reach down to turn on the yaw damper, and of course, there isn't one. You've got to stand on that big rudder to cancel out any snaking."

Bodycombe simulates an outboard-engine failure, and it's not much of a challenge to kick in enough rudder to compensate. "On the type-rating ride, though, we fail two on the same side during an approach," he says. "That's when you've got to be ready to ask the copilot to help on the rudder. You can't be shy about it."

YAF pilot Don Harner has already shown me one feature that I never knew existed: the "formation stick," which is a dead ringer for a sidestick controller, complete with an armrest and a P-51-style handgrip, on the left cockpit sidewall next to the pilot. It's nonfunctional on *Yankee Lady*, and at first I think it's some kind of in-joke (Harner indeed commands a left bank and Bodycombe, in the right seat, surreptitiously feeds in aileron). But no, it's an autopilot controller that was intended to make long periods of close formatting less strenuous. "It worked fine for the lead pilot, but wingmen preferred to hand-fly," Bodycombe said. Scott Smith came up to the cockpit and agreed. "It was too sensitive. It seemed to multiply every movement you made. They thought it would be less tiring for the pilot, but I never heard anybody say anything favorable about it."

Earlier, over lunch at the lovely and exclusive Grosse Ile Yacht Club—hey, when you arrive in a B-17, doors open for you—Harner and Bodycombe had described landing procedures to me. "You don't want to touch the brakes on rollout unless it's really getting away from you," Harner warns. "And it'll do exactly that in a crosswind, with that big tail back there. Land with your heels on the floor and your feet on the bottom of the rudder pedals. As you roll all the way out, sneak your feet up the pedals, and start to use the brakes. But remember, it's easy to flat-spot the tires, and they cost $1,700 apiece."

Harner explains that the Fort "likes to land the way it took off—somewhere between a wheel landing and three-point." If you can think in such precise terms, the technique is to get the airplane into an attitude that puts the extended tailwheel four inches off the ground when the mains touch. "If you try to wheel it on, it thinks it's still flying," Bodycombe says.

Since I'm not going to be allowed to land *Yankee Lady*, I ask Bodycombe if I can at least fly a pattern and approach, and go around at the last minute. He's game. Initial approach speed is about 145 mph with 1,850 rpm and 25 inches. We turn downwind, and Bodycombe drops the gear (which only the copilot can reach). Jackscrews wind the wheels down ever so slowly, but it's reassuring that you can see the whole procedure out the side window. Props come up to about 2,000, still with 25 inches of manifold pressure and 125 mph. Turning base, it's flaps to one-third and slow to 115 mph. Flaps to two-thirds turning final, then full flaps as we roll out of the turn. Twenty inches, 2,300 rpm, looking for 115 mph, crab into the crosswind from the left—piece of cake! I could land this thing. . . .

Everything goes splendidly until Harner looks up from a nap back in the main cabin to see leaves and branches going by out the window. "Ohmygod," he lurches awake, convinced that I have talked Dick into giving me a touch-and-go. No such luck. Power forward . . . flaps to takeoff . . . positive rate . . . gear up, and we're thundering aloft, again a splendid anachronism in the skies over Michigan.

Well, maybe I couldn't have landed it. Harner regains his rightful chair and sits up straight to do battle with what is in fact a stiff quartering crosswind blowing down a 3,700-foot runway. "But look how wide it is," Bodycombe jokes on short final. "Must have been for those Navy pilots." (Grosse Ile was once a U.S. Navy training field.) Harner is working hard, and I begin wondering whether it was a good idea to stand up between the pilots' chairs so I can look over their shoulders. We're using up concrete while Don fights the wind, and soon we're committed, given a B-17's ponderous go-around ability. He hits tailwheel first, but the airplane is stalled enough that there's virtually no bounce. We stop, brakes squealing, at the very end of the runway.

"I wish we had let you make that landing," Harner laughs. "That was no fun."

Was the B-17 overrated, or am I dealing in the oversimplification that comes so easily to someone who wasn't there? "I don't know whether it was or not," muses Scott Smith. "But I do know that I flew 17 missions, and those babies brought me home every time. It's hard to say, but after all, anyone you talk to about that is still alive."

An afterword:

Early on the morning of July 7, 1999, not long after I flew with Don Harner, he was being driven by his wife, Yvonne, to a Ypsilanti, Michigan, hospital after suffering chest pains that almost certainly were the symptoms of a heart attack. At a blinking-red-light intersection, the Harners were rammed by a large truck. Yvonne Harner died instantly. Don Harner survived only a few hours longer but succumbed to massive head injuries. Of all the "younger" pilots that Don had introduced to his beloved B-17, I was one of the last.

Harner was as stout and classic as his airplane, and I was enormously privileged to briefly know them both. ✿

dB Drag Racing

✲

TROY IRVING HAS ONE PRIMO SOUND SYSTEM in his 1985 Dodge Caravan: 72 amplifiers—you got it, seventy-two—and 36 big 16-volt batteries to put out the 130,000 watts of power needed to rumble his nine 15-inch subwoofers. To put that into perspective, by far the most powerful production-car audio is the new $300,000+ Maybach's limousine's 600-watt system. Troy sports $80,000 worth of audio alone. "We need more batteries, but that's all the room we have," he gripes. Cool ride, right?

Not really. At a curb weight of about 10,000 pounds, with virtually no room for a driver (and less for a passenger), the Caravan is basically undriveable. But Troy can sit in a parking space and listen to way-loud tunes, yes? No. His audio system can't play music. It's designed to play a single frequency—74 Hz—wicked loud. Troy is a dB drag racer.

dB (as in decibel) Drag Racing is an obscure but growing international "sport" in which competitors stage side by side and go head-to-head for two or three seconds at a time—hence, the term *drag racing*—to see whose sound system is loudest. (To learn all the details, go to www.dbdragracing.com.) The world record, set by a crew of crazed Finns called Team Loud, currently is 177.8 dB.

The roar of a 747 on takeoff is usually quantified at about 140 decibels, though there's really no way to correlate the wide-spectrum noise of jet engines in open air with a low-frequency pure tone inside a highly reflective enclosure. But since the decibel scale is logarithmic, with every 10 dB increase equivalent to a doubling of sound pressure—otherwise known as noise—these bad boys are creating some seriously loud tones. Another thumb-rule is that all else being equal, every three dB of increased sound from a typical dB Drag Racing system requires a doubling of amplifier power.

Such noise would turn your brain to tofu if it weren't generated into a seriously sealed and uninhabited space—the interior of the car—so competitors in Irving's Extreme class bolt doors shut; Troy also uses industrial jig clamps and a threaded one-inch steel rod and nut through the window. They replace windows and windshields with Plexiglas as much as two inches thick, secure panels with turnbuckles fit for an America's Cup racer, and, in some cases, fill the doors with concrete. Then, while the tone burst is generated, team members lie spread-eagled on the roof and push against the car from the outside to solidify it yet a bit more.

One Extreme competitor in search of ultimate stiffness uses an armored truck, so expect to see an M1A1 Abrams doing a sonic smoky burnout as soon as they're declared surplus.

The sound that leaks out is pretty much what you hear when you inadvertently turn your home stereo on with the volume all the way up and a loose speaker wire: a rattling, destructive, marrow-fluttering hum.

Destructive? Literally. Many teams spend the time between runs repairing blown speaker cones, their equivalent of a real drag racer's melted pistons. "These speakers are like funny-car engines," Troy says. "Some of those cars run for three or four seconds. That's what we design these for—very short bursts of extreme power. Run them down the road for 30 minutes playing music and they'll be useless." At the volumes dB racers run, the temperature of the speaker-voice coils almost instantly goes as high as 500 degrees, and the sound deteriorates.

Competitors must use off-the-shelf, commercially available amplifiers and paper-cone speakers—no boutique exotics allowed—since they're made by the companies that sponsor and support the game. "The speakers we use have to be available to the market, but that doesn't mean they have to be marketable," Troy Irving admits. Some companies manufacture special dB

Drag Racing speakers that they sell to serious competitors for $300 and offer to retail buyers for $2,000 or more, since they really don't want kids rolling down the road trying to play music through them. Some of the setups use so much power that they can empty five or 10 special 16-volt car batteries in minutes.

At the end of each major meet, the four loudest competitors are lined up for the "Deathmatch," a five-minute, winner-take-all face-off in which they fire sound salvos at each other as judiciously yet loudly as possible, trying to make their speakers and power sources survive until time is up. Amid the reek of ozone and hot metal, often just one is left standing. Only heavily sponsored competitors dare play this game, since the cost in equipment is so high.

Many of the cars that staged at a dB Drag Racing event I attended in Toronto were sad-looking beaters, some with zoomy but faded paint jobs advertising their sponsors or the owner's car-audio shop. A Super Street class Nissan Pulsar brush-painted a bilious green putts onto the judging ramp, driven by a kid sitting on a plastic milk crate. Barely visible through the dirty back window is a jumble of amps, cables, and batteries. Yet it blows away its bracket partner with a thunderous 158.2. "Cosmetics aren't going to make it any louder," says Extreme competitor Frankie Valenti.

Valenti is frazzled, having pulled an all-nighter trying to get his GMC van's "enclosure" right. The efficacy of the enclosure is largely determined by the sharp-edged, multifaceted shape inside the van—usually built of wood as much as four inches thick and then fiberglassed—that covers what used to be the dashboard, center console, steering column, and anything else the builder figures will decrease internal volume and direct the sound at the judges' incar microphone.

The shapes are as goofy and angular as a Stealth Fighter's. Some work, some don't. "It's just guesswork," Troy Irving admits. "You start with one thing and if it works, you make the airspace smaller. If that works, you make it smaller again. A lot of it is unquantifiable physics. You're trying to get the wavelength so it matures right at the microphone."

"I could be sitting on a number higher than anybody here has, if I move that back wall forward a foot," says Valenti. "We moved one piece and the level went up 10 dB, but it takes a lot of time and work."

Valenti admits that he's often asked why he pursues this hugely pointless hobby. "Yeah, it's weird. But there are people who have tens of thousands of dollars invested in stamp collections, forgodsake. That to me is weird." He has a point.

Troy Irving's partner Jason Bradley explains: "You start out with a nice stereo in your everyday car, and it grows and grows and eventually gets out of control. The sad thing is, I don't even have a stereo in my daily driver anymore," he laughs. "Every dime goes into this equipment."

Yet Irving's van has traveled barely one mile under its own power during the time he's been competing. "We ran out of gas awhile ago," Bradley admits, "and somebody said, 'How could that be? We just put in five gallons last year.' " (The engine is run solely to charge the batteries.)

"Our competitors can't even spell recession," says dB Drag Racing impresario Wayne Harris. "They're young, some of them still live at home; they put all of their energy and money into their cars. They're competitive. They're at that age."

The competition is taking place in a big convention hall that simultaneously houses a floorful of car-stereo displays. As I wander the exhibit area, the rumble of extreme bass pouring out of Hummers, Escalades, pickups, slammed Civics, and stretch limos is more felt than heard, and I'm almost levitating on a floating carpet of sound. The inevitable silicone-enhanced models in Spandex and low-cut tops have what must be the worst job in the auto-show industry: They are required to sit inside cars trying to look cute while the sound system pounds. Nobody's going to be getting laid tonight . . . "Honey, I've got a headache." It's the audio equivalent of staring at the sun.

Most of the teams wear sponsor-bedecked uniforms and have race-painted and decaled cars. Often there's a "pit crew" of six or eight suited-up mulletheads tending to their car. So it comes as a shock when Jason

Parsons drives onto the ramp in his clean, stock, unmarked '87 Impala and all by himself throws down a 155.8 to win his category in the Super Street class.

"Yeah, I play music through the system," he tells me. "Be silly not to." What a good idea. Does he have any interest in moving up to the Extreme classes? "No, they're out after world records. That and tax write-offs for their audio shops."

Now I get it. ✸

THE SHIPPING NEWS

✿

SHIP-HANDLING ISN'T WHAT IT USED TO BE. When I was a seaman aboard already-ancient Liberty and Victory ships under a variety of flags ages ago, they were 10,000-plus tons of barely controllable iron, steam-powered, single-screw ingots with polyglot half-wits manning the helm—sometimes even me, a 19-year-old who barely knew a hawser from a hoser.

Pushing up a narrow river toward a Gulf port to pick up a cargo of Catholic Relief rice bound for Vietnam—talk about your coals to Newcastle—something went wicked wrong as we approached an open drawbridge, and amid bullhorn shrieks from our Greek captain, the bosun dropped the anchor in a rattle of chain and sparks. It was the only thing we had to stop us, but not before we rammed the bridge and dragged loose an underground power line. It was rush hour, and we spent the remains of the day under the gaze of thousands of parked Cajuns who weren't going anywhere. And when they did get home, they'd find that the electricity was out.

At STAR Center, in Dania, Florida, they train merchant marine officers who conn ships a magnitude and more bigger than those scrofulous tramps, and they do much of that training with enormously sophisticated simulators—the maritime equivalent of the flight simulators that airlines use. Which is appropriate, since a modern ship's bridge is basically a floating glass cockpit, with the same kind of digital displays, joysticks, and power levers that you see when you peek through the cockpit door of a 767 on your way to seat 96G.

The pride of STAR (Simulation, Training, Assessment and Research) is a dead-nuts-accurate replica of the bridge of a typical sophisticated, powerful, maneuverable 21st-century ship. The sim can be configured to handle exactly like anything from a supertanker to a floating-megahotel cruise ship, with a view out the windows to match—360 degrees of computer-generated visuals projected on an enormous screen that entirely surrounds "the ship." Software creates the hydrodynamics that make the vessel respond to winds, currents, waves, decreasing keel-to-bottom distances, channel backwash, and all the other factors that make a huge hull want to turn every way but loose.

STAR has set the simulator up for me to make believe I am aboard a 142,000-ton Royal Caribbean Voyager-class cruiseliner inbound to New York. After the Verrazano Narrows Bridge sweeps slowly overhead, an orange-and-white Coast Guard Dolphin helo curves in across the port bow waiting to drop an inspection team. There's an outbound ro-ro to port, two tankers ahead, and a couple of pesky yachts to starboard, pitching as the white-caps grow in response to somebody downstairs keying commands into the system.

Suddenly the visibility drops to maybe 250 and a quarter. Now what? Oh, good, it's become a moonlit night with a dazzling skyline. And now, a bright, sunshiny day again. The simulator can reproduce everything but motion, though after an hour or two of watching the world bobbing, pitching, and rolling everywhere you look, you actually begin to feel your sea legs.

Inside the bridge, there are two bright radar displays—one the standard rotating-beam, phosphorescent analog screen, and the other a color digital display with our future track generated across the harbor. (It's apparent that we'll eventually take out the Hoboken railroad station if a course change isn't in the cards.) Two data screens display dozens of parameters of the ship's speed, powerplant settings, loading, trim, position, nav frequencies, and, for all I can tell, Broadway ticket prices and availability. The fifth panel is a moving map, a perfect digital replica of a classic hydrographic chart, the yellow land-mass featureless but the sea stippled with soundings, buoys, and beacons.

In the old days, you literally stood your watch, and only the captain sat. Today's bridges have two comfortable chairs, for the bridge officer and the conning officer, and gone is the avast-me-hearties helmsman heaving at a wheel while watching the compass repeater that I'd stared at for endless hours. Course commands are typed and punched into a humming autopilot, and even during close-in docking and maneuvering, a conning officer's palm on the throttling-and-vectoring control does it all.

Mariners come to STAR to train because technology continually advances, and they need to keep up. One of the most important recent advances is podded propulsion.

Typically, big ships have anywhere from one to four screws—propellers—in the stern driven by engines amidships through long shafts. The ship ponderously changes direction in response to rudders flapping and screws pushing forward or pulling aft in a variety of combinations, plus the help of bow thrusters that are essentially sideways screws. Now, however, some of the most modern vessels have done away entirely with rudders and fixed screws and instead have propulsion units that are huge versions of the bottom end of an outboard motor, driven by vertical shafts direct from the engines. They rotate 360 degrees, put out enormous power, and do all the steering as well. But they take brand-new ship-handling techniques, some of them the exact opposite of what a conventional ship requires.

Oh, and don't automatically accept the old saw about a big ship taking miles to stop: while it can easily take 20 minutes to crank a big fixed-shaft diesel from full ahead to full astern, a podded-propulsion cruise ship can do it in several lengths of its own hull.

On a wall of one of STAR's many corridors—there are a number of bridge simulators plus several huge engine-room sims and cargo-unloading trainers, all of which can be linked together in one huge "full-mission" simulator—there's a poster that says, "If you think training's expensive, try an accident." The aerial photo is of two ships well and truly stuck into each other. Airline pilots train on simulators to keep their passengers alive. Mariners do it to literally keep their companies afloat. For the true cost of marine accidents is not lives lost in a quick fireball but endless environmental impact, particularly if the misstep takes place close inshore, as it well might with an island-hopping cruise ship.

The total loss of a liner, which is wildly unlikely, might cost $500 million. The simple holing of a tanker like the huge S/R *Mediterranean*, its photo on another wall nearby, can cost biwyuns and biwyuns, as Carl Sagan would have said.

In fact, it did. The *Med* used to bear the name *Exxon Valdez*. ✿

Eject!

✦

"I'LL SAY IT THREE TIMES," SAID THE pilots who briefed those of us who occasionally got to ride in military airplanes before the era of sequenced command ejections, "and if you're still in the cockpit long enough to hear the third 'eject,' you'll be all by yourself." (Today, a pilot can initiate the ejection of a backseater, or even an entire crew, without saying a word.)

It is from ejection-seat statistics that we learn how dangerous it is to fly military fighters and bombers. One out of every 10 ejection seats manufactured since the end of World War II has at some point been fired for real. The vast majority of those chairs saved lives, and the success rate for the most sophisticated current ejection seats hovers somewhere north of 99.6 percent.

The aircraft ejection seat was independently co-invented by the Luftwaffe and, of all people, the placid Swedes. German pilots made over 60 actual ejections during World War II, while Saab became a pioneer because it built the ugly but powerful J21 fighter, which had a big pusher propeller behind the pilot. This configuration would have made a smorgasbord of any Swede unfortunate enough to bail out conventionally. Designers considered explosive bolts that would have severed the prop or even jettisoned the entire firewall-aft Daimler-Benz V12 powerplant, but decided that to propel the pilot's seat up and clear, by compressed air, was a more elegant solution.

But it was a British company, Martin-Baker, that in 1946 demonstrated the principles of ejection-seat design that still hold true: explode the pilot out of the airplane. What counts, Martin-Baker determined, is the rate of acceleration of the seat rather than the power of the pyrotechnics that launch it. A small bang rapidly applied to the pilot's fanny, as though he were simply an artillery projectile, is far more harmful than is a bigger one progressively dispensed. Martin-Baker proved that 21Gs of peak acceleration for no more than one-tenth of a second was the most the human spine could routinely absorb, and that the rate of acceleration could not be greater than 300Gs per second. In fact, manufacturers long ago decided that 150Gs per second, thank you very much, was a safer standard.

A live human was used for proof of concept: the legendary Benny Lynch, a lowly Martin-Baker mechanic who was awarded the British Empire Medal in 1948 for his quite voluntary daring. Lynch was persuaded to do the original ground, and then, flight tests of what generations of U.S. Navy pilots came to ruefully call the Martin-Baker Backbreaker. So called because in the days before rocket-motored seats, the Air Force used a sophisticated double-base charge that was smoother than the Navy's triple-bang British catapult. It took a while for pilots to become accustomed to sitting atop what were essentially small cannon shells. (The most dangerous way to eject is accidentally, on the ground. Unfortunate crew chiefs working on a cockpit, and even ditzy children who found their way into jets at air shows, have inadvertently ejected themselves from parked airplanes. Without a parachute, the ride back down from 200 feet is invariably fatal.)

What does it feel like to eject? Ross Detwiler, today a corporate pilot for Citigroup, captaining a Challenger and a Falcon (without ejection-seats), was 24 and had been flying North American F-100 Super Sabres over Vietnam for a year without a scratch, "so I felt blessed," he says. "I'm gonna get away with it, I thought."

He didn't. Leading a four-ship strike against a truck park in Laos in April 1969, Detwiler's Hun got stitched by a quad 23mm as he released two of his four 750-pound bombs.

"My Number Two, in trail behind me, yelled, 'You're on fire—bail out.' But that was a losing proposition, because they didn't take prisoners where we were. So I headed east toward the sea. Climbing through about 13,000, the fire went out, because the burning hydraulic fluid was gone. Number Three came up on my wing to look me over, and next thing I know, he's peeling off to get out of there and calling, 'You're on fire again!'

"I cleaned up the cockpit, put my kneeboard in its compartment, sat myself up straight, pulled the ejection handle, and the canopy went off. The noise scared me, and I thought I'd been hit again. Oh, and the funny thing about the F-100 was that the five-step ejection sequence was printed on the canopy. Step one was to 'assume the position.' Step two was to raise the ejection handle, which blew the canopy, and steps three, four, and five went away with the canopy.

"I remembered to squeeze the ejection trigger, under the handle, but nothing happened. I squeezed it again and was about to do it a third time when the instrument panel started to move down in front of me. Then the top of the windscreen bow went past, and then I'm looking down into an empty cockpit. Up on the nose of the airplane, I could see the big, luminous 13 that the pilot who 'owned' the airplane had put on it, giving fate the finger, and I remember thinking, 'Yeah, and I'm picking up the tab on this airplane.'

"So now I'm sitting about five feet above the airplane, and I remembered a friend who'd ejected twice telling me, Man, when you come up above canopy level, you really tumble, and I'm thinking, Gee, I'm not tumbling at all. And then suddenly I tumbled backward twice, real fast, and I thought whoa, there it goes.

"I'm waiting for the seat belt to open, thinking I'll open it manually if it doesn't, I've got plenty of time. I look down at my lap and see just a puff of white smoke, and both sides of the seat belt come apart.

"It was like in the movies, when everything happens in slow motion. I always thought that was just a special effect, but that's exactly the way it happened with me. I don't remember at all being fired out. When the charge in the seat [an M39 20mm shell] went off, I never felt it.

"That's about how long it seemed in my mind—about the length of time it takes to tell it. But my Number Three said he'd heard me transmit, 'I'm gonna get out,' and when the T in *out* was pronounced, the canopy and I came out of the airplane at the same time."

Detwiler ended up hanging 15 feet off the ground, cocooned in his para-chute canopy upside down. "It took the paramedic hanging from the Jolly Green 11 minutes to cut me loose. I grabbed him and got onto his jungle pen-etrator, and he got me into the helicopter and laying down on the floor. Then he said, 'Ah, you can let me go now, Lieutenant.' I was pretty hopped up.

"When we got back to Da Nang, my knees were pretty well bruised from flailing during the ejection—I'd forgotten to tighten my leg straps—and they asked me if I could walk or wanted to be carried to the ambulance. First, I said I'll walk, but a Pan Am 707 had just landed, and here were seven or eight girls from the cabin crew coming over to see the rescued hero fighter pilot, so I said, 'Gee, why don't you carry me, guys?' "

The best ejection seats in the world today are Martin-Baker's Mark 16E, being fitted to the F-35 Joint Strike Fighter; the U.S. ACES III seat in the F-22 Raptor; and the latest Russian Zvezda K-36 system. The latter has been seen by air show crowds in close action, as Russian pilots punched out—at hospitality-suite height—of fighters that were about to become lawn darts, most spectacularly a MiG-29 at the Paris Air Show in 1989, and again at Paris, this time a Sukhoi 30, in 1999.

These are enormously complex handmade machines, with a parts count typically of over 1,300. Essentially, the seat is an aircraft, even to the extent of containing its own life-support system for high-altitude ops. "It is a flight ve-hicle," says Business-Development Director Mike Annen of Goodrich Univer-sal Propulsion Systems, which manufactures the ACES seat. "It doesn't fly for long, but it must be aerodynamically stable, and it has its own engine. It has to come out in a variety of conditions and actually fly before you get man/seat separation. It's an aircraft within an aircraft."

Current ejection-seat technology is considerably more complex than the early jack-in-the-box designs, in which jettisoning the canopy, initiating the ejection, getting loose from the seat, and pulling the ripcord were all done manually. Today, snugging straps and netting immediately tighten to keep a pilot's limbs from flailing, and a chemical charge initially urges the seat out of the airplane. (The next generation of Goodrich seats will have head- and neck-support airbags as well.) Then a rocket motor takes over, firing in response to a microprocessor's decisions about the pilot's weight, the attitude of the airplane on ejection, and the airspeed and altitude at which ejection took place. On some seats, a gyro-stabilized vernier rocket provides directional control. At higher speeds and altitudes, only a stabilizing drogue parachute will deploy, while at low altitudes the main chute is fired immediately, and the seat is kicked free by a seat/pilot separator motor.

"The ACES III is good from zero to 700 knots, from the surface to 50,000 feet," says Annen. For a marketer, he's being modest: one F-16 pilot is known to have launched a predecessor ACES II seat not merely at sea level, but while in a steep descent of thousands of feet per minute, after a pitch-control failure. His wingman saw the huge splash of the airplane hitting the water, out of which soared like a phoenix the rocket-powered seat.

Still, ejection is not a routine activity. When all four crewmen aboard a B-1 that crashed in the Indian Ocean in December 2001 ejected during an aborted bombing mission during the Gulf War, offensive-systems officer Lieutenant John Proietti had been a six-footer when the B-1 took off. He was five-foot-eleven when he hit the water. ✿

Ambulance Driver

THE MOST INSULTING THING YOU CAN CALL a paramedic or an EMT is "ambulance driver." They despise the term, for drivers are the bottommost link of the medical food chain, dangling directly below the people who rinse out bedpans.

I never wanted to be a doctor but always wanted my very own siren, so I'm proud to say I *am* an ambulance driver. For 36 hours a week, I wear a pager and an EMS uniform while I stand watches as a volunteer driver for COVAC—the Cornwall Volunteer Ambulance Corps—in a placid upstate New York town. The control knob in fact reads, SIREN/WAIL/HYPERWAIL, plus a button that blows the WONK-WONK air horn. I'll see your cop-car and raise you three strobes, though I'll fold in the face of a fire engine.

Dedicated ambulances like the three gaudy, strobe-bedecked but top-heavy lumps that I drive are a surprisingly recent phenomenon, dating back only to the late 1960s. Until then, some hospitals had in-house ambulance squads typically using what the trade called "Cadillac station wagons"—check out Claire's pistachio ride in *Six Feet Under*—while others called on morticians to do double duty with their hearses.

That's when a few companies started to build modular ambulances—a purpose-built box for medical equipment, a stretcher, and attendants atop the frame rails of a standard light truck. Surprisingly, the business hasn't changed much since then. The quality of the box and the crash safety it affords for the EMTs working back there has increased dramatically, but it's still a rolling cubicle packed with cabinets, drawers, cubbies, and compartments filled with basic emergency-medicine equipment, plus a portable stretcher and the greatest possible space to work on its unfortunate occupant.

Crash safety is no small thing. There is very little training for ambulance drivers, and many are either well-meaning volunteers or ill-paid professional EMTs or paramedics with no relevant driving experience, and they are in command of an ill-handling truck and a siren that automatically adds 15 mph to their adrenaline-fed throttle foot. The safety record for ambulances is dreadful, and there is some justification for the claim that the most dangerous part of a motor-vehicle accident is the ride to the hospital.

I learned some of this when I traveled to Goshen, Indiana, to see COVAC's newest piece of equipment: the $100,000 rig I'd be driving in a few months, coming down the assembly line at Medtec Ambulance. It looked like a bread truck, the poor thing—the module shiny, unpainted aluminum covered with crayon scrawls of trimming instructions, the cab still plain refrigerator-white.

Good ambulances are handmade, and the amount of work that's going into a job like COVAC's new rig is impressive. In the case of Medtec, much of it is done by local Amish—bearded men doing precision cabinetry while bonneted, plain-shifted women assemble circuit boards and wiring harnesses. "The Amish have great mechanical aptitude," Medtec GM Tim McDonald says. "These are people who live in homes without electricity and make it work. And when something breaks, they don't just run down to the store; they fix it."

The module is a double-walled structure as joisted, studded, and braced as a small house, hand-welded entirely of aluminum, a nasty metal to weld without distortion or burn-through. A huge harness runs inside the roof panels to feed the ambulance's exterior and interior lights—a good module can be made almost as bright as an operating room—as well as pumps, chargers, defibrillators, and radios.

Every seam is sealed, and the rubbery floor laps up the sides where it joins the cabinetry like the tiled deck of an Australian pub designed to be hosed out after an evening of what the Aussies call chundering. Bloodborne pathogens are one of an EMS worker's biggest fears, and a good ambulance has nowhere for bodily fluids to hide. (One of an ambulance driver's least-favorite duties is swabbing out and disinfecting the rig after a gory run.)

Security is newly important too. Even COVAC, operating out of Mayberry, USA, services one nearby small-city hospital where our rig must be physically guarded outside the ER entrance lest the rolling pharmacopeia be pilfered by the stoned locals. At an MVA—a motor-vehicle accident—an ambulance is typically unattended while everybody, including the driver, works to extricate and stabilize victims, and now we're hearing that terrorists are looking for ambulances to use as bomb trucks. The best rigs have complex central-locking systems and, if they carry paramedics, lockable drug drawers where the professionals can stash their serious meds.

Everything fits atop a Ford chassis, since the Blue Oval seems to be the only company that cares about the tiny (6,000 new units a year) ambulance trade. Ford installs the huge alternators that ambulances need, and routes heating and a/c lines aft for easy connection to the module, so with few exceptions, you'll rarely see a Chevy or Dodge ambulance.

Ours will be an E350 dualie. It will also have an odd snow-chain system that we'll need amid the small but steep snowbelt mountains hereabouts, but that in my truck-deprived life I've never seen before. I remember watching my father laboriously wrap chains around the tires of our '52 Chevy with cold-deadened fingers before snow tires were invented, but now, at the flip of a switch in the ambulance's cab, two motors that spin pinwheels of short, straight chains like crazed ninja devices drop down so that the chains are rapidly flung in an endless whirl under the rear tires.

Commercial ambulance services that need to make a per-run profit use ordinary Ford vans with raised roofs, which can be bought from Medtec for as little as $45,000. The community volunteer corps that charge nothing for their services but can make a fervent pitch for funds to the town board or city council are usually awarded enough money to buy a good box-on-frame rig, which typically costs from $80,000 to $110,000.

But the Big Dogs are the ADs—"additional-duty" ambulances. Many local fire departments have less to do than in the past, as stiffer building codes and wiring standards make residential fires infrequent. A community ambulance might answer ten calls for every fire-engine run, so the fire dogs, looking for business, are increasingly getting into the EMS act. They're ordering up buff, outa-my-way "rescue trucks"—mini-Macks smaller than a fire engine but *way* bigger than an ambulance. These ADs typically go for $125,000 to $150,000, "but we've built them for as much as $190,000 when they're loaded," McDonald says. "Custom compartments, power seats for the EMTs, custom mounts for extrication tools, a water tank and pump for car fires, a $15,000 generator just to power floodlights, an on-board cascade oxygen-refill system, TV cameras for backing up *and* so the driver can watch what's going on in the module . . . the fire guys love their trucks."

Hey, I can dream, can't I? ✸

FLING-WINGS

✻

A COUPLE OF YEARS AGO, I FLEW a big ex-Vietnam Marine helicopter, a 1,525-horsepower, radial-engine UH-34D that in its day was the Hummer of the air. In truth, I simply manipulated the controls from the copilot's seat for 15 minutes once the owner of the nicely restored Dog had taken off and established us in level flight over Houston.

I promptly disestablished level flight. After a few minutes of watching me corkscrew through the Texas sky, the pilot keyed his mike and said, "If I have a heart attack, you have 40 seconds to live."

He was right. I'd been flying a wide variety of airplanes for 30 years, but the only way I'd have gotten the porky Sikorsky onto the ground was as a pile of parts. Yet *trying* to fly it was the most fun I'd had in the air since I soloed a Stearman. I needed more.

My friend Mario, an FAA examiner and helicopter instructor, operates three piston and two turbine helos off his broad lawn in . . . well, I can't tell you, because Mario keeps a low profile and takes new students only if they've been signed off by the Pope. Most are aerial-unit cops, state troopers, medevac pilots, and corporate guys. And now, me, terrified to be in the hands of this hardass who'd seen it all.

"Flight instruction is supposed to be fun," Mario growls as we preflight a two-seat Robinson R22 for my first lesson. "But learning to fly a helicopter is sheer torture. I can't *tell* you how frustrating it is. If there were a way for me to plug you in so that I could instantly transfer what I know, I'd do it. But flying a helicopter requires repetition. Doing it over and over again until you finally memorize the right control inputs." My first few hours of instruction will be miserable, frustrating, and terminally discouraging, Mario promises me.

"A professional pilot with thousands of hours of business-jet time will come to me and beg, 'What am I doing wrong; how can I figure out how to tame this little thing?'" Mario says. "All I can tell them is, 'Give me more money. You haven't done it enough yet.'"

The R22 costs $175 an hour with Mario instructing. This is getting close to twice what it costs for flying lessons in the equivalent fixed-wing Cessna. Mario's Bell 206 JetRanger goes for $600 an hour.

The controls of an airplane are intuitive. Push to nose down, pull to nose up, turn left, turn right. You can put your feet flat on the floor, ignore an airplane's rudder pedals, and still get where you're going. (I once numbly flew a 200-mile trip in a Beech Bonanza that I'd never flown before, with the rudder locks still in place.) Airplane controls also require gross inputs. Most lightplanes, from two-seat Cessnas to big pressurized twins, are about as touchy as a '49 Hudson.

A helo's controls are different. Until the movements become second nature, you have to *think* about what you're doing. One cliché image is that flying a helicopter is like patting your head while rubbing your tummy. Another compares it to balancing atop a greased beach ball. And the controls require only pressure, not real movement. A helicopter-pilot friend who is also a serious radio-control modeler swears that a Robinson trainer requires tinier control inputs than do the inch-long joysticks on his r/c controller.

Mario is a muscular little fireplug of a guy, but his movements become as sleek as a tango dancer's as he plays air guitar with imaginary helicopter controls to show me how the aircraft responds to the merest thought transmitted through the proper pressures. "Women make better students," he says, "because they naturally have more finesse. If you need to wear your wife's underwear to get in touch with your feminine side, do it."

There's a stick called a *collective* in your left hand. It goes up and down like a sports car's parking brake and has a motorcycle-grip twist throttle on the end. In the era of that old UH-34D I'd mistreated, you turned the grip left to add power, right to remove it. Today, you do that only to warm up the engine and run a magneto check. Even on my tiny R22 trainer, the engine rpms are managed automatically, and power is actually added by pulling up on the collective, which coarsens the pitch of the main rotor blades and forces the engine to put out more power.

In your right hand, you're holding another stick called the *cyclic*. Say SIGH-click. (And while we're at it, don't ever say chopper or copter. The only people who do are TV blow-drys and other amateurs. If four syllables are a burden, call it a HEE-low.) With the cyclic, you control the helicopter's bank, the up-or-down angle of its fuselage, and generally, your track over the ground.

Ignore a helo's anti-torque pedals at your peril. Ignore them, in fact, and you'll find yourself a passenger on a pinwheel. They control the small propeller—or in some cases, a ducted fan or blower—at the tail of the bird. If it weren't there, a helicopter's fuselage would rotate in one direction around the main rotor shaft about as fast as the rotor blades were turning in the opposite direction. Particularly while trying to hover, the pedals are crucial.

Bomber pilots call it CEP—Circular Error Probable—the circle around a target within which 50 percent of their bombs are statistically expected to land. It's also useful in determining the amount of real estate a novice helo driver will cover while attempting to hover over a spot. My CEP during that first lesson was half an acre of the onion fields and grass-sod farms over which we practiced, to the bemusement of at least one tractor driver who obviously had no idea that my CEP happened to include him.

The tractor guy was gone during my next session with Mario over the same fields, but he'd have been impressed. I actually hovered. Not for long, but for those few pure, perfect, I'm-a-hummingbird moments when the little Robinson rested calmly in the air three feet above the vegetables, I was in control of the most marvelous flying machine the world has ever seen. Thank you, Igor Sikorsky.

But one question, Igor: How did *you* learn to hover on that first-ever helicopter flight without Mario to calmly say, "Left pedal, more left pedal, watch your attitude, add some power, left cyclic, keep the nose up, left some more . . . I've got it." ✿

The Concept-Car Concept

✲

THE WORLD'S FIRST CONCEPT CAR, TRADITION HAS it, was the General Motors Y-Job, a boat-tailed, two-seat, experimental Buick roadster that appeared in 1940. Fact is, the Y-Job was created not to be displayed at car shows but so GM design chief Harley Earl could drive around Detroit in the coolest ride in town.

In the six and a half decades since the Y-Job—they were then called dream cars—concept cars have served a variety of masters other than Earl. Some have been serious technology demonstrators, others playtime projects for designers temporarily freed from the drudgery of designing door handles. Some have been opinion-samplers to test public reaction to a new design direction, others little more than dressed-up prototypes intended to say to auto-show crowds, "Get used to it; this is the sedan we'll be selling you in two years."

And some—Harley Earl would have understood—have been cruise missiles aimed straight at the competition's penthouse office suite. "There's a lot of pecker-matching going on between executives at car companies," says automotive-design consultant Robert Cumberford, "and that's been true forever."

"There's an arms-race quality to it today," says David Laituri, the principal industrial designer at the Massachusetts firm, Design Continuum, the largest privately held industrial design-and-development group in the U.S., numbering both BMW and GM among its clients. "It's blinding, the sheer number of concept cars today. It's the price of entry. You have to do something."

Sometimes "something" is simply next year's pickup truck, but with 26-inch wheels, rubber-band tires, a 15-liter V12 from the company's earthmover division, and a $40,000 iridescent paint job—the answer to the question, "Jeez, what are we gonna do for the Detroit show?" Sometimes concept-car designers think so far outside the box that reality becomes a distraction. At the 1996 British Auto Show, a design team from Coventry University displayed Concept 2096, a wheel-less, window-less, driverless vehicle that looked much like something you might find sucking your thigh after you'd waded through a toxic Cambodian swamp. The powerplant? "Slug Drive," which the creators admitted hadn't yet been invented. Hey, no problem.

One of the more outrageous of recent concept "cars" was the 500-hp Dodge Tomahawk V10 motorcycle introduced at the Detroit show in 2003. As a usable vehicle, it was essentially worthless. "I disagree," says Design Continuum designer Alan Mudd. "What it does is show the public that Chrysler is trying things, that Chrysler is made up of a bunch of passionate people. Even if it is a total adolescent wet dream, it has value because it tells the consumer that Chrysler is full of excited, creative people who just want to try some great new stuff."

"We would never do that at Volvo," says, not surprisingly, the general manager of Volvo's Concept Center, in Camarillo, California, Lars Erik Lundin. "It has to do with a company's core values, so there's absolutely no risk that Volvo would ever do such craziness."

Lundin also points out the potential downside of too-imaginative concept cars. "You can raise people's expectations. Sometimes the things they see don't happen. We had a hybrid concept car in 1992. We still don't have a production hybrid, yet we showed that it was in the future 12 years ago."

"There was a Range Rover concept called the Range Stormer at a recent Detroit show," Robert Cumberford recalls, "and it provided Land Rover with a serious problem. Everybody loved it, but there was a traditional Range Rover coming out that wasn't as zoomy as that, and they were suddenly in a bit of a panic, realizing they didn't have a car like the Range Stormer to sell. If you make a concept car something you can't carry through with, you disappoint people. You don't want to do that."

In the past, when high tech was visible, comprehensible, and demonstrable, concept cars were the new-technology demonstrators of choice. Dimming and then automatic headlights, power windows, cruise control, rear-view TV, parking sensors, remote locking, keyless ignition, LED brake lights, variable-opacity roof glass, voice-activated in-car phones, on-screen traffic-information displays, heated seats, rain sensors, folding metal roofs, and dozens of other now-common features first appeared on auto-show concept cars as early as the 1950s. Increasingly, however, concept cars have become marketing tools and styling exercises rather than tech testbeds. (The interesting exception is concepts intended to represent a single huge technological leap, like GM's skateboard-with-a-body fuel-cell Hy-Wire platform, or the Toyota Volta 155-mph high-performance hybrid.)

"Yes, it is difficult to make technology visible," Anne Asensio agrees. (A Frenchwoman, she is GM's executive director for Advanced Design.) "The designers say okay, here we have an opportunity to start from scratch again. The cell phone did a good job of that. It changed telephone design totally. We have to break the code of the way we used to do things, and that's what concept cars can do."

Asensio feels that "an innovation simply for the sake of innovation is nothing. You must have something that brings benefits for people. Something that people can immediately connect to and say, 'This is going to make my life better, or more interesting, or safer, whatever.'"

"It works if the technology can be made understandable," says Alan Mudd. "One of the reasons hybrid technology is so successful is that it's fairly easy to understand. People understand an electric motor plus a gasoline engine; they don't need to know all about regenerative braking. But hydrogen and fuel cells—it's so much more difficult to grasp that technology."

Says Mudd's cohort David Laituri, "It's getting harder to present thinking-outside-the-box stuff on concept cars. It's hard for the public to stand 20 feet away from a concept car on a stage and get a technological story. It's an emotional story you're getting." After all, BMW's baffling iDrive, introduced on the 1999 Z9 "technology testbed," must have seemed like a good idea at the time.

Actual future-tech is usually introduced and assessed on modified production cars that are indistinguishable from what's currently in the Toyota or Ford dealer's showroom. They are eminently driveable but in fact contain hybrid drivelines, sensor-crammed automatic accident-avoidance systems, steer-by-wire controls, or other magnum-leap advances.

Where technology and concept cars do increasingly intersect today is in the gap between fantasy and reality, between the original one-off handbuilt show car and a mass-production vehicle. Where once clay models had to be hand-measured with calipers and rules, tooling laboriously built and prototypes assembled and reassembled until everything fit, now computer-aided design and manufacturing (combined with malleable new materials and apparently endlessly adaptable platforms) means that today's wacky concept—a convertible pickup truck, a mock mini-Hummer, a bubblecar—can be tomorrow's production vehicle. Concepts that light enough people's fires are going into production as what some are calling OUVs—occasional-use vehicles—niche-car toys. Sell 10,000 a year and you've made back the nut.

"Yes, it's easy to go from concept to production," GM's Anne Asensio agrees. "It's also very smart. You have a lot of new market segmentation. People are saying, 'I want a car that belongs just to me, responds to me, reflects my lifestyle and values.' So with this explosion of multiple niche products, it's so important to do a great show car and make sure you hit the ball. Concept cars are good tryouts. We learn a lot by doing them."

Concept cars run the gamut from engineless fiberglass shells that have to be moved by forklift to finely crafted, handmade but driveable metal machines. Sometimes, cars presented as concepts are even made from production tooling with working drivetrains borrowed from another model, but these are really what Robert Cumberford calls "teaser cars"—preproduction previews.

They're driveable within strict limits, of course; recently, Cadillac invited a group of us to sample its enormous 1,000-hp, 13.6-liter, V16-engine Sixteen sedan. It was a flop, for the car broke down, an affair reported in amused detail by *The New York Times.* ("When the guests arrived, the Sixteen was sitting on jackstands with an oil slick expanding beneath it . . .")

There was a time when most concept cars were straight out of *The Jet-sons*—dream machines with phony jet exhausts, rocket fins, fighter-plane bubble canopies, and Madonna-bra front ends that were far-out visions of the future. "The cars of the GM Motorama era were idealistic," says David Laituri. "Everyone knew these weren't cars you were going to be able to buy, but they enjoyed their being presented as 'dreams.'"

Appearance and reality, however, were two different things. As Walter Boyne wrote in *Power Behind the Wheel*, "the dream cars' greatest contribution . . . was in their reverse validation of the venerable principle that form follows function. As the dream cars were without function, they could have any form. Pointlessly long hoods, swooping fenders, skirted front wheels, and improbable driving components were carelessly conferred upon cars that would never travel a distance greater than the radius of a turning platform at an auto show."

"The dream cars of the '50s were myths," adds Robert Cumberford, "to make people dream about the future that was going to be so glorious in the aftermath of World War II." Yet one of the more spectacular of the 1953 Motorama dream cars was really one of Cumberford's "teaser cars," he points out, "because it was already committed to production and was offered for sale in June of that year." It was called the Corvette.

Concept cars have in fact often been nothing more than gaudily jeweled prototypes already committed to production. They are shown in order to prepare the marketplace for a new design direction, a styling leap that needs to be approached just a bit gingerly. Many car enthusiasts assume that manufacturers build concepts to assess the reaction of media folk and car-show attendees, but megacompanies are rarely that cautious or uncommitted. Typically, they're a generation or two ahead of the auto journalists who award their stamps of approval to the concepts.

When Chrysler showed the precedential Portofino large sedan concept at Frankfurt in 1987, the company was already committed to the cab-forward era it presaged. When Ford introduced the sharp-edged GT90 at the Detroit show in 1995, during the heyday of jellybean cars, it simply made public the New

Edge creased-lines styling that they were already applying to Mercury Cougar and Ford Focus prototypes.

Lars Erik Lundin of Volvo reveals that there have even been cases of companies axing a project and then showing it as a concept car. "Some failed production-car projects will make it into the concept-car arena because you have the vehicle anyway, and you want to see what the reaction would have been if you hadn't decided to kill it. That I know has happened," Lundin says.

Lundin and other professionals are skilled enough to be able to tell when a concept car is just a design exercise and when it's a stalking horse for an upcoming production vehicle. "I look for how the details have been made," Lundin explains. "If they're from production tools, then you know it's probably something that's going to come out in a year or two. If you check the details, it's an enabler for me to tell whether the company will probably come out with that car or if it's just a test to see what the reaction is going to be."

When Porsche showed the Boxster concept car at the 1993 Detroit Show, it knew that it had to create an entry-level sports car to help save the ailing company, but it really didn't know what it would look like. American designer Grant Larson's Boxster concept was met with such universal enthusiasm that Porsche froze the design a month later and had the car on the market three years later.

Though Larson's Boxster concept was a remarkable example of near-seamless translation from auto-show jewel to showroom model, concept-car designers often make this journey problematic. "What you can afford to do with a concept car is ignore the legal demands, like bumper regulations, emissions regulations, pedestrian-safety ratings, and all kinds of things," Lars Erik Lundin points out. "So you may be disappointed when you go from concept car to real car, because the design suffers."

"One of the problems of bringing a concept car to production is that's when it hits the design-by-committee level," says Alan Mudd. "That's when it stops being the singular vision of a courageous designer and has to go through the miles of red tape that a huge company will impose upon it. That's when the excitement can get sucked out of it."

Almost without exception, modern concept cars are designed not by salarymen sitting in Toyota City or cubicled workers staring at monitors in Detroit but by free-thinking stylists in California, Barcelona, Milan, Paris, and other vibrant culture centers. Not only because the sun is bright but because the suits are a thousand miles away.

"If you're in the middle of headquarters," says Lars Erik Lundin, "your best designers are constantly pulled into fixing production projects. They're never in peace and quiet long enough to make a concept car. Tomorrow's work is always more important than the future." Lundin works in Southern California nearly 9,000 miles away from Volvo's dour headquarters in Gothenburg, Sweden.

But there's more than mere distance. "You could be working in Dearborn in a satellite studio without a single executive for 20 miles and you still wouldn't feel as good about your life as you would doing the same job in Pasadena," says Alan Mudd. "That kind of environment has an incredible impact on designers."

During its very brief useful life—after all, nothing is deader than last year's spectacular show car—a concept car could be considered priceless, the product of endless hours of handwork, of artisanry seen nowhere else in the auto industry. For a concept car is inevitably one of a kind and irreplaceable. It was a disaster for Chrysler when two spectacular show cars painstakingly crafted for the company by the Italian firm Ghia ended up on the bottom of the Atlantic off Nantucket in 1956, deep in the hold of the liner, *Andrea Doria.*

The Chrysler Ghias are still sleeping with the fishes, but don't waste millions trying to retrieve them. Typically, a truly important concept car is worth $200,000 to $300,000 on the collector market—a price range identical to that of good Mercedes-Benz 300SL Gullwings.

Occasionally, however, auction fever inflames bidders. In June 2002, Christie's hammered away 51 Ford concept cars and prototypes that the company had for some reason decided "to put . . . back in the hands of the people," as Design VP J Mays put it. It was one of the only times in decades that a manufacturer had released any such vehicles, and most went for an average of under $50,000 apiece. Star of the show was a 1992 Ghia-built Focus concept roadster, the primer and bodywork beginning to show through the paint, the leather seats scratched and water-stained. Christie's guessed it was worth $100,000. It sold for $1,107,500, the current concept-car record.

Ultimately, concept cars are the height of automotive impracticality. They range from rolling jokes to out-of-nowhere design breakthroughs. Some have taken us down such cul-de-sacs as intelligent-highway automatic cars sniffing buried cables like a beagle on the scent; others have presaged the engineering leaps that may finally sever our oil dependency. Some have been arrogantly ugly, others so gorgeous you just know the automotive future is bright.

But after all, what would a car show be without concepts? Nothing more than the world's biggest dealer showroom, and what's the fun in that? ✿

HAWK

✦

I BUILT AND FOR NINE YEARS FLEW an airplane called a Falco, which is Italian for hawk. But it wasn't until May of 2002 that I flew a *real* hawk—a cold-eyed, scimitar-beaked, red-brown Harris's hawk that perched on my gloved left hand, flapped off into the Vermont air, dove at mice and voles like an F/A-18 with bin Laden in the crosshairs, and eventually returned softly to my hand. Okay, it didn't return to my *hand*, it returned to the small cube of raw beef that I'd stuck between thumb and forefinger of my fist.

Yes, this book is titled *Man and Machine*, but hang on—a hunting hawk *is* nothing more than a brainless killing machine with the aerodynamics of a Reno racer. It is also hard, fast, and shiny, and I'll offer a few further rationalizations presently.

How fast? Well, how's 247 mph in a dive, a peregrine falcon's speed verified in recent tests in California sponsored by the National Geographic Society? There is a piece of film shot in the 1940s that was rigorously analyzed by the British Royal Navy, which concluded that it revealed a hawk diving at 273 mph, but skepticism abounds. Still, I can't think of a faster animal on the planet. Even in level flight, peregrine falcons average about 40 to 50 mph.

A diving hawk looks like a very angry top-gun Tomcat at full-aft wing sweep. The truly fast ones only take birds in midair, since they'd crater if they dove on a ground animal. The Harris's with which I practiced was better-adapted to surface targets. Feathered Stukas, hawks have tiny tabs called alulas at about the midpoint of their wings that form little leading-edge slots with which they can vary the direction and speed of a dive. Hawks also have tiny bony protrusions in their nostrils that act a bit like the splitters inside a supersonic jet's intake. They keep the airflow in a dive from rupturing the bird's air sacs.

Even a hawk's bill is mission-specific. It is slightly notched—"toothed"— on each side so that it will fit just between a rodent's cervical vertebrae for the terminator bite.

Hawks don't sing, soar for fun, preen, or socialize at bird feeders. Their only vocalizations are a squawking "pick *me*, pick *me*" when you enter the rookery where a bunch of hunting hawks are waiting to be taken out, and something that sounds like Yoda ruminating when they're on your fist and sense live meat somewhere nearby. A small raptor has to eat 20 to 25 percent of its body weight per day—the equivalent of a 200-pound running back putting away a daily 40 to 50 pounds of Big Macs and fries. All that hawks think about is food. They spend about 90 percent of their life standing motionless on a perch, either digesting what they've eaten or looking for more.

A hawk can see a mouse a quarter-mile away, if it moves, and they can see about three times as far and as clearly as can a 20/20 human. Their "flicker vision"—the number of discrete images they process per second—is said to be four times that of the human eye/brain combination. Perhaps because a hawk's pair of eyes weighs *more* than its brain. We have about 200,000 cells per square millimeter of the image-acquiring parts of our eyes. Hawks have a million or more. A raptor's eyes are huge. The eyes of an eagle, for example, are the size of yours and mine, yet an eagle has one-twentieth a human's body mass.

An Alaskan eagle can kill a wolf. "Anything that can kill a wolf can maim a man," writes Stephen Bodio in *A Rage for Falcons*. "A female eagle may weigh 13 pounds and be able to exert hundreds of pounds of pressure at the points of her talons."

The key to using hawks for sport-hunting is not that the birds kill—that's what they do for a living—but that they return to you after doing so. It's not because they form a bond with their handler, for you have as much chance of turning a hawk into a pet as you do of having a shark fetch your slippers. The trick is fuel management. No free-flight modeler would launch a model airplane fat with gas, for that would let the thing fly so far they'd never get it back. In the same manner, hawks are only flown by falconers when their "tanks" are a quarter full, or perhaps even on reserve.

A falconer knows to the ounce the empty weight, as a pilot would say, of his or her bird. If a Harris's hawk is just an ounce or two heavier than that, it's good to go: it'll run out of gas before flying too far and will always be forced to refuel—to return for the chunk of meat on your fist.

As a new falconer with a bird on the fist, your first surprise is that the vicious-looking beak a foot from your face isn't a danger. A hawk's main weapons are its talons, not its beak, and it would no more think of biting than a pit bull would consider kicking you in the shins. The second surprise is the bird's weight: the size of a pullet, it feels as heavy as you imagine a robin might, for its bones are quite hollow—a tube-frame fuselage, in effect.

When you "cast" the bird by simultaneously stepping off with your right foot and urging the bird into flight with your left hand, it's like launching a delicate balsa-and-Mylar model. You want to be smooth, not wrist-snappingly harsh, and your heart flies with the bird just as it would with the model. To fly a hawk is to *be* a hawk.

Even better is when the hawk returns from a perch. A cowboy whistle usually brings it back, and the bird wastes not an erg of effort on its nicely stabilized descent. On final, it goes right down into ground effect, less than half a wingspan off the ground, and adjusts its outer feathers like outspread fingers—quintuple-slotted flaps and ailerons combined, in effect. Coming over the fence, again as a pilot would imagine it, the bird brakes delicately, feeds in some aft stick and flares, gear down and locked, bleeding off speed by climbing the final five feet from ground level to fist.

It's no accident that so many raptor terms have become part of aviation's language, whether as the name of my Falco or that of the Spitfire's Merlin engine. (No, it had nothing to do with the Arthurian magician. A merlin is a kind of hawk, and that nomenclature was preceded by V12s called Kestrels and Eagles.) A pilot doesn't care a fig for the flight of a robin or a grackle, unless his turbine is about to ingest it, but a hawk invariably makes any aviator stop and look. A hawk is a living airplane. A seagull, despite Richard Bach's best efforts, runs a poor second.

The sport of falconry is 4,000 years old, and except for the occasional use by African nobles of cheetahs to chase down game, is the only example of wild creatures being used by humans for hunting. The falconry equipment used today—leather jesses to hold the hawk's legs, an intricately sewn helmet-hood to cover its eyes while traveling, a swung lure that mimics the hawk's prey, the thick glove—is much the same as what Kublai Khan used when he rode forth with a staff of 10,000 falconers on horseback tending his 500 hungry raptors.

Except for the tracking beacon; today, valuable hawks aren't flown until a tiny radio transmitter trailing a thin antenna wire is strapped to one leg, so the bird can be found if it gobbles up enough mouse meat to change its mind about the need to go home. It has a range of 20 or 30 miles in flat, treeless terrain but only two or three miles in hilly, forested Vermont.

What in the Middle Ages was the nobility's equivalent of r/c model airplanes has today become, if not radio-controlled, at least radio-located. ✸

Tanks, Hot Rods, and Salt

✸

WORLD WAR II LEFT IN ITS WAKE precarious peace, Levittown, the Berlin Corridor, the GI Bill, Harry Truman, and serious hot-rodding. Hot-rodding?

"Sure," says Pomona, California, rod restorer Pete Chapouris. "You take a guy who had some kind of Ford or Chevy that he'd been working on and driving to high school, and you throw him into this big airplane. All of a sudden he looks around and goes, 'Wow, man, that braided stainless-steel oil line, that works. And what about this steerin' wheel, this is plenty cool. And look at these neat switches and this bitchin' bucket seat. And these seat belts—man, I could keep from flyin' out of my car.'

"World War II spawned hot-rod racing just as going to the moon spawned computers," Chapouris points out. "Same sort of technology shift, just in a different time frame. Clevises, jam nuts, Dzuses, Allen-head bolts— you look at the early postwar rods and they're just covered with aircraft stuff."

In fact, some of the cars themselves *were* aircraft stuff. They were called "belly tank lakesters," invariably shortened to "tanks." A tank was a starkly simple, mid-engine, open-wheel raindrop of a car, a soapbox racer on steroids. It was made by dropping a driver and an engine into a World War II–surplus aircraft drop tank—the disposable, streamlined, aluminum fuel containers carried under the bellies and wings of fighters to supplement their internal tankage.

"Here was something the government had designed to have a low coefficient of drag at 300 mph in flight," says Dennis Varni, who races a tank on the

Bonneville Salt Flats in Utah even today—one of three true lakesters still running. "It was the perfect thing to cut a hole in so you could poke your head out and see, put your flathead Ford engine in it, and go. It was affordable racing. The body was already built."

The first tanks were somewhat primitive devices, really little more than a pair of frame rails narrowed to fit inside a belly tank and bolted or welded to a solid 1930s rear axle. Front suspension was minimal. But at a time when "roadsters"—most of them the classic bucket-bodied 1932 Fords—had top speeds of 116 to 126 mph, the stunning reduction in frontal area afforded by the World War II belly tanks immediately kicked speeds up to 150 and 160 mph. Still, the tank drivers wanted more.

How fast would the cars go? Well, in the late 1940s and early '50s, while Mercedes-Benz, Maserati, Alfa Romeo, and even Indy 500 teams were running fancy overhead-cam engines that powered monster racecars to maybe 150, the kids running on California's dry lakebeds and Utah's salt pans were tickling 200 mph with flathead Ford V8s—an antique engine (introduced in 1932) that resembled nothing so much as two Model A four-cylinders sharing a common crank. And by the 1960s, with the arrival of enormously powerful supercharged Chrysler Hemi engines, one tank had done 292.

Tanks, in fact, were the fastest open-wheel racecars on earth. And in their compact dimensions and "backward" driver-in-front configuration, they presaged the teardrop-shaped, rear-engine European grand prix cars that would revolutionize racing in the early 1960s. This was a coincidence, not the product of either enlightened hot-rod engineering or fat California wallets. "Most of us had no money at all," recalls ex-racer Tom Sparks. "Not only no money, but no real facilities or equipment. Oh, we could put a V8 engine in a Model A frame, but that was about the end of our capabilities."

But ingenuity they had in excess. "LA was the phonograph-record headquarters of the country," Sparks says, "and the various companies would cut their records on oversize aluminum master discs. Well, those masters would fit over a standard 16-inch wheel pretty good. I had a friend who worked at

Decca, so we'd use them to cover both the inside and outside of the wheel, reduce the drag." (One can only hope they were masters of discarded Johnnie Ray takes and not forever-lost Sinatra sessions.)

A Navy veteran named Bill Burke is entirely responsible for the hankering for tanks. Retired today but still playing with hot rods, including a 1937 Ford and a record-setting Studebaker Avanti, Burke recalls, "I saw a group of wing tanks on a barge, being taken ashore at Guadalcanal, and I thought, *My god, what a beautiful piece of streamlining that is!* I got a tape measure and went aboard and measured one of them. I knew the dimensions of a Ford rear end and the size of an engine block, and I could see they'd fit."

Burke made his first lakester out of a small P-51 auxiliary tank, and following convention, he put the engine in front. "There wasn't enough room to put it in back and still get a seat and your legs in front of it, so I sat upright behind it and stuck out of the tank a long ways. Yeah, it was a helluva windy ride," he laughs, "but it did the job."

After that, Burke switched to 13-foot-long, 315-gallon Lockheed P-38 drop tanks—a streamlined shape exactly three feet in diameter at its fattest point. "You'd laugh at what I used for frame tubing inside those tanks," he says. "I used old Model T rails for the first one, but then I switched to PBY Catalina wing strut tubing for the frame rails. It was clean, strong, good-looking, and pretty good-size."

The first surplus drop tanks were available for as little as $35 or $40 apiece, and hundreds of them were piled in surplus yards. Racing hobbyists weren't the only ones interested in buying them, though. Ranchers split the tanks in half for use as animal watering troughs, and many were also used as houseboat pontoons and advertising display attention-getters.

The tanks consisted of top and bottom halves bolted together along horizontal flanges. The aluminum shells didn't themselves hold the fuel; they contained a seamless rubber bladder. Since the top half was encumbered with the fuel opening and all the hardware necessary to fasten the drop tank to the airplane, standard procedure was to build a car out of two smooth, mirror-image bottom halves.

"You open up an aerodynamics textbook and there it is, the classic teardrop described as the optimum streamline shape," said Mark Dees, a California lawyer, former tank racer, and amateur auto racing historian who died in a traffic accident in the late '90s. "Well, that shouldn't be a surprise. Those tanks were designed by Lockheed, and they could open a textbook too."

Nor should it be a surprise that postwar California hot-rodders were attracted to things aeronautical since a number of them were Air Force vets (excluding Navy man Bill Burke). "When you look at the early photographs taken on the dry lakes, there are a lot of guys wearing big sheepskin jackets and leather ball caps with the bills turned up," Pete Chapouris points out. "Most of them had been in the Air Corps—gunners, mechanics, fliers. Imagine being 19 and flying a P-51 and the next day you're back in Glendale with a major need for speed. You can't afford to buy an airplane, so you go racin'."

Alex Xydias, perhaps the best known of all lakester drivers, was one of them. (In 1952 he booted the *So-Cal Special* to 198.34 mph, faster than any normally aspirated flathead Ford-powered vehicle has ever traveled.) "The Air Force had a great need for mechanics, and most of us who had a chance to volunteer rather than be drafted, we signed up to be airplane mechanics," remembers Xydias. "Gosh, we saw guys from Southern California wherever we went."

Some of these vets, in an attempt to reduce the frontal area of their lakesters even further, used the considerably smaller 165-gallon teardrops intended for P-51s and P-47s. This severely compromised the aerodynamics, however, by putting both the driver and the V8 engine's cylinder heads and carburetors well out into the considerable breeze. Even the full-size tanks were a packaging problem. Says Dennis Varni, whose car is based on a P-38 tank, "In ours, you lay down inside the tank. Your head isn't exposed. You look between your legs through a very narrow, very small windshield that's about four feet in front of you, so it's like looking through a tunnel." What's it like at that speed? Varni laughs. "It's a great high," he says. "Better than sex: It lasts longer."

Mark Dees got around the packaging problem by building a twin-boom racer: two tanks connected by the axles. He outfitted one tank with the engine and seated himself in the other. "It was a fun car to drive," he remembered. "You're sitting in the left boom and you look over to the right and you'd swear that somebody in another tank has driven right up next to you at 260 mph."

"The thing I remember most about driving a tank was not the speed but the incredible racket," muses Xydias, an ex-B-17 flight engineer and top-turret gunner. "You were down inside this echo chamber, and the straight-cut gears in the Halibrand rear end were howling and the engine was just screaming. The salt was never as smooth as it looked, so you were kind of bouncing along, everything making such a racket that it was just unbelievable." Xydias once described it to a friend as "riding in a 55-gallon oil drum with three or four guys running alongside hitting it with baseball bats."

The noisiness of the lakesters wasn't the only drawback. "I would expect all those tanks were pretty horrible to drive, in terms of handling," opines Kirk White, a New Smyrna Beach, Florida, dealer of vintage and exotic cars who is in large part responsible for a recent surge of interest in classic hot rods as collector cars and museum pieces. "I suspect it took a great deal of nerve to pilot one of those things at anything approaching 200 mph. Aerodynamically, there was nothing to press them onto the ground the faster they went [as happens with a modern racecar]. They were good for the time, but I'll bet that driving one today would terrify the best race driver in the world."

The golden age of the tank peaked with Xydias's spectacular 1952 demonstration of what a nitromethane-fueled, prewar flathead engine could do. But his 198-mph achievement was soon overshadowed by the beginning of a new era—that of the postwar, overhead-valve V8. These simple but sophisticated engines, particularly the Chrysler Hemi and the ultimately ubiquitous small-block Chevy, stoked America's love affair with the automobile. With such horsepower available, fickle need-for-speed hot-rodders quickly moved to plumb the possibilities of envelope bodies handcrafted from sheet aluminum—full streamliners—thus eliminating the enormous drag of the tank's unfendered wheels.

Still, a surprising number of the old tank racecars remain—in one form or another. Not that there were many to begin with; probably no more than 20 that utilized the classic P-38 drop tanks. But they changed owners, engines, and colors so often that for a while they seemed to be everywhere. ("Yeah, but remember, you're talking about dry-lakes racing, a sport that was done by maybe 300 or 400 people total," Dennis Varni points out. "So 20 cars would be a lot.")

Today, a war-surplus aircraft belly tank would be snapped up for thousands of dollars by a collector, a warbird enthusiast, or an entrepreneur eager to restore or replicate a classic lakester. "I would say the old tanks are starting to become, not priceless, but certainly up in the $150,000-plus range," says Pete Chapouris, who was hired to restore the *So-Cal Special,* which today sits in the Peterson Automotive Museum in Los Angeles. "The *So-Cal* car, with its history, is going to be worth $200,000 before long."

Indeed, the old cars have been validated with that most quintessentially Californian of honors: Today you can buy replica fiberglass drop tank half shells and build your own tank. ✿

Vee is for V12

✿

Sixty years ago, the fastest airplanes on the planet were powered by enormous, complex V12 piston engines made by Rolls-Royce, Allison, and Daimler-Benz. Sixty years later, some of these very same engines are still running, powering weekend warbirds, aviation-museum artifacts, and Reno racers. Fewer than 10 men in the U.S., however, have the knowledge, skills, equipment, and temperament to keep them flying. Meet some of them.

Junkyard Cats

East of Gilroy, California, on a two-lane running from Nowhere to Nevermind, amid sere brown hills and stubbly fields, down an unmarked, cactus-bordered dirt road is Dwight Thorn's company, Mystery Aire Ltd.

The mystery is how this collection of ramshackle, oily industrial sheds turns out the most powerful, most reliable, and most admired air-racing V12 Merlins in the world. Engine blocks and parts are everywhere. Scarred junkyard cats sun themselves atop pallets of superchargers. Cylinder heads are stacked like cordwood, the man-high layers alternating direction as though they were laid out to season in the breeze. Every upturned sump and valve cover is filled with eucalyptus leaves and spider nests. Crankcases are slowly sinking into the sandy soil—ashes to ashes, aluminum to aluminum.

But make no mistake—Dwight Thorn builds awesome engines. They routinely win Reno races. Looking a bit like Wilford Brimley in bib overalls, the white-haired, 66-year-old Thorn laughs when I say that most people expect Merlin magicians to wear white shop coats. "Overalls are the most functional garment ever made."

Thorn is putting the finishing touches on a bright red Merlin with mirror-polished aluminum valve covers that will soon fill the snout of the two-seat P-51D Crazy Horse, which flies as a trainer and warbird-wannabe thrill-rider out of Kissimmee, Florida. Seventy-five percent of his work is overhauls of stock engines such as this one. "But the two or three racing engines we do every year take just as long as all the others put together." Thorn charges $60,000 to $80,000 for a stock overhaul, depending on the condition of the run-out engine, and $160,000 to $180,000—and up—for a labor-intensive, 3,800-hp full-race motor.

Exactly what do you do to hop up a Merlin? "Simple," Dwight smiles. "Disconnect the boost [limiting] control. We've seen 150 inches of boost, which is where the gauge stops. And which is probably just as well."

Most of us accustomed to more conventional motor sports—CART, Nextel Cup, Formula 1—assume that "tuning" plays a large part in separating the prime V12 builders from the also-rans. That porting and polishing, boring and stroking, bench-flowing and blueprinting, tinkering with timing and tuning exhaust pipes must be a large part of successful air-race engine-building.

Nope.

The category in which the V12 engines run at Reno is called "Unlimited," and that's exactly what it is. The rules basically say the engines must reciprocate and turn props. There is no displacement limit or ban on any performance-enhancing device such as turbocharging, supercharging, turbocharging plus supercharging, nitrous oxide, designer fuel, exotic materials, or extreme weight-saving techniques.

As a result, the top V12 builders put their engines together with the emphasis on using the strongest possible parts, the best possible reinforcing of weak areas (such as the relatively vulnerable crankcase of the Merlin), the most careful assembly and torquing, and stock profiles and settings for the camshafts, valves, and ignition.

And then they turn up the boost. The more air and fuel that the blowers can cram into the engine, the more horsepower it makes—and the stronger the engine must be to withstand the unholy pressures inside the combustion chamber.

Thorn's specialty is adapting beefy, never-run Allison connecting rods to Merlin crankshafts and pistons to replace fatigued old Rolls-Royce rods. This allows the engine to operate safely at 135 inches of supercharger pressure but at lower rpm. Before Thorn's imaginative fix, racing Merlins turned as much as 3,800 rpm, which meant the propeller was spinning inefficiently fast, the tips going supersonic. Now racers can back the revs down to 3,300 or 3,400, allowing the prop to get a better bite on the air but sending piston pressures into the stratosphere.

Most of Mystery Aire's client's aren't racers. "We're dealing with a different kind of customer now," Thorn admits. "Back in the 1960s and '70s, the majority of the owners worked on their airplanes, had military experience—some had even flown the P-51 in the service. Today, it's the nouveau riche. They're like the Ferrari guys—people who've bought something they assume will appreciate in value, and in five years, they'll go on to something entirely different."

Between Rolls-Royce, Packard, and Ford of England, 165,000 Merlin engines were made during and after World War II—more than any other single engine, including the ubiquitous American radials. Today, enough usable parts survive worldwide to make perhaps a few thousand. Warbirders weep when they think of the several acres of Los Angeles land that in the '60s were carpeted with surplus Merlins and Allisons. A speculator had bought the engines for pennies per pound, but when the real estate became vastly more valuable than the engines, they were shipped to Japan and melted to make beer cans.

"There were commercial applications that used up these engines, too," Thorn recalls. "Thousands were used as mud-pumping engines in oil fields. They'd run them on natural gas straight from the wells, and when they stopped, you'd throw them away. It was cheap horsepower—an engine you could buy for $200 that would put out 1,000 hp round the clock."

Thorn builds his best engines with what the cognoscenti call "transport banks." Between 1948 and '50, Rolls-Royce made the strongest and most durable Merlins of all for Canadair, which used them to re-engine DC-4s that it called Northstars. These 1,760-hp Merlins were made to pound away hour after hour without missing a beat, and they made use of everything Rolls had learned about building durable V12s. Their cylinder banks and heads are the Merlin gold standard, and if you want to build a racer, you need them.

What about nitrous oxide? The Luftwaffe used it to augment their simple, single-stage superchargers, and hot-rodders inject it into the engines of everything from Civics to Camaros for instant acceleration. (NOX is a powerful oxidizer and "thickens" the air—and therefore the amount of fuel—that an engine inhales.) Thorn will provide nitrous but says, "It's hard to carry enough to make it worthwhile. A hot-rodder can fit a five-gallon tank and go play all night, but with an engine this size, you've got to have a lot on board. Nitrous was real popular for boat-racing Merlins and Allisons, where you needed acceleration out of each corner, but in air racing, you're at top speed when you hit the starting line, and you stay there through the race."

Scattered and stacked throughout Thorn's warren of shops are shelves, boxes, racks, and pallets of Merlin parts, many still in their original, sealed Rolls or Packard packaging. "I've been able to buy a couple of complete [shop] inventories over the years, so I have a good stock to work from," Dwight says. "I could probably build 20 complete engines from scratch. Not counting the things that wear out, like bearings, I probably have 200 engines' worth. But someday, there will be one little widget that nobody has anymore, and you won't be able to finish an engine unless somebody steps up to the plate and manufactures it."

Part of the problem is that "A Merlin has six times as many pieces as an Allison," Thorn points out. "I blame it on socialism. The more parts they had to make, the more hours of labor were needed, and the more make-work the government had achieved."

Thorn's protégé Mike Barrow builds his own engines alongside Dwight and pitches in to help Dwight when needed. When Thorn retires—he's 66—it's likely that Barrow will take over the business. "I had a cousin, Louis Norley, who was an ace with the Fourth Fighter Group," Mike muses. "He flew with Gentile and Blakeslee, and I've always had a thing about P-51s and Merlins. It's neat to be able to work with this stuff, and I like the air racing, too. I've been a crew chief—though when you're both the crew chief and the engine guy, no matter what breaks you're in trouble," he laughs.

"People my age—I'm 38—when I tell them that I overhaul Rolls-Royce V12s for a living, they don't know what I'm talking about."

THE ODD COUPLE

Sam Torvik and Bill Moja have worked together for 30 years and still argue about whether the shop radio is too loud. Torvik is a small, trim-bearded, tightly wound Merlin specialist. Moja is a big, shuffling, mustachioed galoot, the kind of man whose pants ride low and shirttail is usually out, and he prefers Allisons. "The English engines . . ." he shakes his head, "full of lousy rubber seals and way too many pieces. They're like Jaguars burning out at the side of the road all the time. I don't know why we're still doing Merlins, there's so much labor in them."

Torvik and Moja are the V12 builders at JRS Enterprises, in a shabby-looking brick building in suburban Minneapolis, in the shadow of Chevy and Ford dealerships and just down the hill from a housing development. The "R" has fallen off the JRS sign, and a constant stream of traffic rumbles past on a four-lane. It was once largely the hobby shop of racer John Sandberg, who was killed in 1991 in the crash of his remarkable homebuilt Unlimited racer, *Tsunami*, a tiny mini-Mustang that remains the smallest airplane ever to have carried a Merlin. Today, JRS Enterprises is basically a fabrication shop filling small aerospace-industry machining contracts, but in the back room, engine-building continues almost as though nobody really knows how to stop it.

"Why are we still doing this?" Moja says. "Because we always have. Nobody's making much of a living doing this stuff, because it's just for rich boys and their toys—that and flying museums. But it's warm in here during a Minnesota winter, and you get to go home at 3:30."

Torvik is happiest left to himself, so that's just fine with him. He has his own small engine-assembly area, where he's finishing up an early single-stage-blower Merlin that will go to England for Stephen Gray's Fighter Collection. (Early Merlins and Allisons are far rarer than later, more powerful, and more sophisticated variants.) "I don't know what it's going into, either a Spitfire or a Hurricane," Torvik says, "but Gray flies the hell out of everything. He puts the hammer to it."

"And the weight of the hammer is directly proportional to the depth of the pocket," Moja chuckles. "He breaks 'em, we fix 'em. We've got customers who figure that if they get 100 hours out of an engine, they're doing good. All depends on the depth of the pocket."

Torvik is impressed by what he's seen of German World War II engines, having recently worked on a BMW radial from a Focke-Wulf FW-190. "Their technology was so far ahead of ours at the time, it was easy to see," he says.

Moja demurs, of course. "They were way too complicated. You didn't have to be a rocket scientist to work on an Allison. They were made to be serviced by 18-year-old kids."

Moja is building an Allison for a P-40 restoration in the next room, amid piles of engine parts and tools. Hanging on a wall nearby are a huge Merlin conrod bent a good 10 degrees from straight and a supercharger impeller that looks as though somebody has punished every blade with a mason's hammer. They're from *Tsunami*'s Merlin the day Steve Hinton brought the power up and the anti-detonation injection failed. "If the ADI system fails, you can't reach anything in the cockpit fast enough to keep the engine from blowing up," Mystery Aire's Mike Barrow had said to me, and the display proved it. The explosion nearly blew *Tsunami*'s cowling off.

JRS does 15 or so engines a year, most of them big radials for commercial operators—firebombers, heavy-lift helicopter operators, and the like—as well as restorers. They do only one or two V12s a year, but are usually at work on several at a time, while waiting for overdue supplies or missing parts. "The commercial stuff, that's a push, because those people need their engines," Moja says. "A V12, the worst that happens is a rich boy misses an air show. We haven't worked on a weekend since September '91." Which, as it happens, is when Sandberg died and JRS was out of the air-racing business.

"The round motors are probably more reliable than the V12s," Moja admits, "but remember, the V12s were made for an entirely different purpose. They were the truck engines, hauling bombs for the most part. The V12s were the hot rods, made to go balls-out all the time. You're asking me to fly behind it? I'll take the radial every time."

A variety of ailments can afflict a V12 when it's asked to do too much—even Moja's Allisons. They're prone to cylinder-liner distortion if overboosted, because the liners are locked to the block both top and bottom, and when uneven expansion is exacerbated by sudden overheating, the liners deflect slightly and let the combustion charge sneak past the rings, which is invariably fatal to that piston, which then destroys the head with its shrapnel. "The Merlin's liners, even though they're thinner, stay pretty much round, because they float at the bottom end, where they're sealed by O-rings," Moja points out.

Imagine a soup can with both its top and bottom cut out, and you have a small, very thin cylinder liner. Grab it by each end and twist, and it distorts—becomes slightly ovalized. Grab it by only one end and you can't do that.

"See that semi out there?" Moja asks, pointing at a battered white trailer outside the shop. "It's filled with cylinder heads that prove you can't overboost an Allison." Sandberg began his short air-racing career with an Allison-powered P-63 Kingcobra. An inveterate tinkerer, he continually had its engine modifed in a variety of ways, never leaving well enough alone. "We modifed and modified that engine and kept blowing it up," Moja says. "Then we finally took it back to stock and it ran better than ever."

"There are no tricks to building a good engine," Sam Torvik avers. "Everybody thinks there's magic involved, but there isn't. You just have to build it right and use the best parts. It's becoming harder and harder to find them, though. When you keep blowing these things up racing them, where are you going to get parts for your regular customers?"

Some of the simplest yet most difficult-to-find parts—new crank and rod bearings, for example—would be easy for a competent fabricator to manufacture, "but the people who could do it don't want the liability," Torvik points out. "If a Rolls-Royce part fails, it fails. If a JRS part fails, they sue us. It would have to be done by an overseas company [sheltered from liability]. I keep hearing rumors that there's some foreign company gearing up to build complete new dash-seven Merlins. Yeah, there will be enough of a market for something like that if people keep dragging these airplanes out of the bushes and rebuilding the airframes."

Torvik and Moja blame "the boat people" for the dearth of parts. "When the hydroplane guys found out they could use Merlins and Allisons [for Gold Cup racing], it drove the value of a core from $250 to $25,000," Moja says. "And then the tractor-pullers came along. Then when they screwed up all the engines they could find, they went to turbines. If I get a call from a tractor-puller looking for parts, I won't even talk to him."

Most of a Merlin or Allison engine—"the core"—never wears out, unless the ravages of time simply corrode it beyond redemption or racing use over-stresses it. The crankcase and crankshaft, cylinder banks, accessory-case and valve covers, heads and cams, supercharger, prop-reduction gears, and various pumps and fittings are usually salvageable. During a rebuild, new pistons, rings, and bearings go into the core, and other parts such as cylinder liners and valves are reground and freshened.

Unless a buyer has fallen for the bargain price of a former hydroplane engine, the main clue being nonstandard fittings for oil-scavenge pumps at the aft-facing end of each head, since the engines sit at a considerable angle in boat-racing use. Those engines have lived a life of very high rpms with the prop jumping in and out of the water and their cores are useless as anything but . . . well, boat anchors.

Of course, the air racers destroy engines too. "Yeah, but they only do it once a year," Torvik smiles.

STEALTH SHOP

Tehachapi, California, is a small, high-desert town, but when I ask for directions to Vintage V12s, nobody knows what I'm talking about. Mike Nixon, a scholarly, preoccupied-looking man who wouldn't be out of place wandering across the Caltech campus, likes it that way. "I don't do any advertising, and I let the local paper do a story on us once every four years as long as they don't print where we are. It would only attract the tire-kickers."

At one point while I'm in Nixon's compulsively neat shop, a deliveryman from town shows up and takes in the spectacle of a dozen or more glossy Merlins and Allisons in a spectrum of M&M's colors. "What are they for?" he asks, wide-eyed. For airplanes like those in the photos and paintings that cover the walls, Nixon explains. "You mean for, like, hobbyists?" Well, something like that.

Nixon's "hobbyists" are, for the most part, serious restorers rather than racers. "I can do a restoration engine and see it come back for an overhaul in six or seven years," he says. "Racers fly your engine for two or three years and blow it up. See that yellow supercharger and set of valve covers?" he asks, pointing to a rack of Merlin parts. "They're from an engine I first worked on in 1978, and it's on its fourth owner since then. I've been working on the engine in Bob Hoover's Mustang for 28 years."

Mike admits that he can't hand-pick his customers, but some he does steer clear of. He thinks of the wealthy athlete who bought a P-51 and called to request an overhaul. "I was on the phone for 15 minutes and couldn't get a word in edgewise," he recalls. "I hung up and said, 'He's dead in a month.' I was right: he flew into a hill while doing a low-level inverted pass."

He says, "I don't like doing a project unless the customer comes out here and sees the shop. I like to meet him and see what he envisions at the end of the tunnel. An Oshkosh warbird trophy? Just flying for fun? High performance? Maybe even going to the Reno races? Most of the dot-com guys don't have the patience to listen to what I want to say, and I don't have the patience to deal with them."

Nixon was until recently a full-race-engine specialist, but he has burned out on the serious competition. Besides, he says, "It's far better for us to have four or five guys who fly our engines in the Silver and Bronze races at Reno [largely populated by basically stock, authentic warbirds], have a great time, and tell everybody about it than it would be for us to win the Gold. Or, worse, be leading the Gold and scatter an engine. It takes years to get over something like that."

Nixon knows that well. "We peaked in 1982, when we built the engines for four of the seven finalists at Reno and our engine in *Dago Red* won. We were overwhelmed with work after that," he says. But such reputations, if not easy come, are certainly easy go. Several Nixon race engines blew during a subsequent season, due to problems he traced back to a piston-ring supplier, and the gossip mill began to grind.

"The only Gold racer I'd have any interest in now would be a Griffon-powered airplane, because it would be a challenge and because there's so much Griffon stuff available," Nixon admits.

Many warbird enthusiasts accept as gospel that the Merlin was a spin-off of Rolls-Royce's Type R racing engine, which powered the famous Supermarine Schneider Cup floatplanes. In fact, the 1,650 cu. in. Merlin was a derivation of the 1927 Kestrel V12, and it was the 2,240 cu. in. Griffon that actually was the production version of the same-displacement R. Development of the Griffon was under way as early as 1933, but was then put aside because the aborning Hurricane and Spitfire needed a smaller, lighter engine.

When the V1 Buzzbomb attacks on England became considerably more than an annoyance, Rolls shoehorned Griffons into what became enormously fast, low-level Spitfires designed to run down V1s. After the war, Griffons became the engines for the Avro Shackleton maritime patrol bomber. Hundreds survive, having led a life largely of low-power, low-level loitering. There are even Griffon "box engines" available, still in crates after being overhauled by Rolls-Royce.

Vintage V12s has accumulated a considerable stock of Griffon parts, but what Nixon is proudest of is their selection of "early-engine stuff." With almost 150 P-51Ds flying and a considerable selection of late-model Spitfires, restorers are today embarking on more interesting, unique projects. And if you want to do an Allison-engine A-36 Mustang, a long-nose P-40, or a late-'30s Spitfire, you may well need to come to Nixon for the right parts. He guesses that his trove's value is at least "a couple million," but who can put a

value on racks of prop reduction gears that look big enough to fit a ship engine, a box of thousands of tiny lock-tab washers in a Whitworth size that no
longer exists, or a drawer filled with irreplaceable valve springs?

Nixon's most recent project has been the total restoration of a rare Daimler-Benz DB 601 V12 for a New Zealand collector's Battle of Britain–era
Messerschmitt Bf 109E. "The biggest problem has been all the magnesium
parts—intake manifolds, valve covers, accessory cases, things like that," he says.
"Since they're all down at the bottom of the engine when it's mounted in the
inverted position, moisture gets at them and they corrode away. If the engine
is from a wreck that was sitting belly-down in a field somewhere, forget it."

Nixon points to the German engine's original valve covers, amid a shelf of
equally useless DB 601 parts. They are magnesium doilies, filigreed with rot,
which is why it took parts from three donor 601s to complete the job. At that,
Nixon had to have a brand-new prop reduction-gear cover, a casting about the
size and shape of a bedpan, manufactured. "Pattern, casting, and machine work,
it cost $20,000," he says. "I look at that and laugh when people suggest build-
ing an entire new Merlin. It would cost $1 million per engine, easy."

The rebuilt DB 601 will cost its owner nearly $300,000, plus the $100,000
he paid for the original core and the parts engines, for Nixon has put a year
and a half into the job. It's the second 601 he's done; the first one took three
and a half years.

"The German, American, and British V12s are fairly similar in general, but
the compexity of the DB 601 is obvious," Nixon says. "The British and Americans did more in-the-field maintenance, whereas the Germans would just
send the whole engine back to the factory. They could change the engine in
a Messerschmitt in a little over an hour." Which is why you see World War II
photos of shirtless, oil-covered GIs with backward baseball caps standing on
sawhorses while they pull cylinders and replace pistons. The Germans left that
work to men in white shop coats.

"Still, there were very few people either at Rolls, Allison, or Daimler-Benz
who knew the whole engine," Nixon claims. "Almost everybody was a specialist. It was only in the 1950s and '60s that we evolved to generalists who

actually work on the whole engine. Guys like Dwight Thorn and me and a few others who have basically had to learn the whole engine."

Is Nixon's business growing as warbirding increasingly becomes the sport of kings? "Stable is a better word for it. We've had some huge incremental increases, like when in the late '70s a lot of ex-South American airplanes became available, and then in the late '80s when all the Spitfire gate guardians came down to be made flyable for the 50th anniversary of the Battle of Britain, but I don't think there are any more 'secret' warbirds out there anymore. The legendary 50 P-51s that were supposedly in China turned out to be nonexistent. When the Berlin Wall came down, that was the last time a large group of World War II aircraft suddenly became available.

"But the nice thing for us is that when you restore an airplane, you never see it again. When you restore an engine, it comes back for an overhaul every six or 10 years." ✿

WALLY 118

✿

IF YOU LIVE ANYWHERE NEAR WATER OR WATCH *Miami Vice* reruns, you've seen go-fasts—those garish, arrogant, pencil-hulled runabouts, big V8s bellowing through open pipes. What you probably haven't seen, and wouldn't believe if it did blast past your porthole, is a Wally 118. You haven't seen it because there's only one, and it's in the Mediterranean. And you wouldn't believe it because this narrow, angular, black-glasshoused superluxury go-fast is 118 feet long, looks like a Stealth Fighter that missed the runway, and is squirted to a top speed of nearly 70 mph by a total of 16,800 gas-turbine horsepower. Price: $24,410,000.

To put into perspective that much grunt in a 110-ton boat designed to accommodate just eight people plus a four-person crew: Awhile ago, I found myself aboard a fast, modern Danish car ferry crossing the Kattegat, the body of water between Denmark and Sweden. That substantial jet-drive, tunnel-hull ship displaced 500 tons empty and had not much more than twice the Wally 118's 16,800 horses, but they were plenty to propel a cargo of 200 cars and trucks, 800 passengers, a restaurant, and a busy bar at a very impressive 50 mph.

The Italian company Wally Yachts (the name is a reference to an obscure Hanna-Barbera cartoon character, Wally Gator, as though Boeing had named itself Snoopy Group) has since 1993 been a high-tech sailing-yacht innovator. The Wally 118 is their first foray into mega-powerboats. All of their boats have composite hulls, the Wally 118's reinforced with carbon fiber and flanked by two huge air intakes reminiscent of the Lamborghini Diablo's angular, extendable airboxes.

The Wally 118 actually has five engines—two 370-hp Cummins diesels plus the three 5,600-shaft horsepower Vericor TF50 gas turbines, derived from the Lycoming T55 that powered the big Vietnam-era Boeing Chinook helicopter. Almost half of the boat's entire interior space is taken up by the powerplants, leaving just three staterooms and minimal crew quarters.

The diesels are for low-speed maneuvering, since even at idle, the turbines put out enough power to make waltzing around a marina challenging. The diesels run the impellers for the two steerable outboard waterjets. Once free of the dockyard, the turbines take over and a third, fixed central jet drive joins the raging chorus.

Most mega-yachts are furnished like the Dennis Kozlowski Suite at the Bellagio, but the Wally 118, though superbly crafted of exotic materials in and out, with thin wood veneers over honeycomb cores and lots of exposed black carbon fiber, is surprisingly spare—more Italian industrialist's office than tycoon's playpen. "This is really just a big dayboat," says Rhode Island yacht designer Roger Marshall. "It's not something you're going to live aboard for three weeks while you motor across the Atlantic; it's a boat you're going to drive as fast as you possibly can with 16 bimbos on board."

Not only does a Wally 118 need substantial berthing space, but its 5,812-gallon fuel tanks would suck your local marina dry. (The TF-50s burn marine diesel fuel.) Miles per gallon? Actually, it's gallons per mile—about 14, or to put it another way, 951 gallons per hour at top speed.

You'll also need a professional crew and should budget at least $400,000 a year for that. One of them will have to be a marine engineer/mechanic with turbine experience. If a Cigarette or a Donzi is the Turbo Mooney of fast boats, the Wally 118 is the Learjet, filled with complex (and dangerous if mishandled) hydraulic, electrical, and electronic systems that, like a bizjet's, need to be thoroughly understood by pros prepared for fast action. "If one of those engines comes apart, it'll go right through the bottom of the boat," says Greg Mullen, publisher of *Dockwalk*, a magazine for professional yacht crewpeople and captains.

Wally plans to build further 118s on order only. "I don't think it's a marketable product," says Roger Marshall. "It's a high-speed boat that is way overpowered and expensive as hell to run, so the market consists of very few people."

Yachting magazine technical editor Dudley Dawson, who sailed aboard a Wally 118 for a demo run, points out that though several different gas-turbine luxo-yachts have been built recently, "almost without exception, the second owner of those boats, and the second boat in the series, have diesel engines rather than turbines. Very few people keep the turbines. They're finicky to maintain in a marine environment, and the fuel consumption—gallons per horsepower—is much higher than that of diesel engines."

Still, there is demand out there for mega-yachts—the ultimate realization of the classic definition of a pleasure boat: a hole in the water into which you pour money. Last I looked, there were two 400-footers under construction in Germany, and a 500-foot private yacht was also known to be quietly abuilding elsewhere in Europe. (Which is longer than the tramp freighters and tankers aboard which I crewed during my misspent youth.)

"You start talking that kind of money, there's speed freaks who have it too," says Greg Mullen. "But after the first [Wally 118] is out there, and maybe the second, that's probably it. The thing about mega-yachts is that the next buyer always wants to outdo the guy before him. The next boat always has to take it a little bit further, a little bit faster."

If so, the increments of advance will be minute, for water is an obstinate medium. Imagine where we'd be if today's cars and airplanes went not even twice as fast as their century-ago equivalents. In 1902, a boat named *Arrow* did 45 mph, a world record. Not a raceboat but a narrow, classic Edwardian steam yacht with a deckful of passenger chairs and a foursquare wheelhouse, she was 132 feet long and developed 8,000 hp from two steam engines. The similarly proportioned Wally 118 has doubled the power but not nearly doubled the speed. ✿

B-17 *Yankee Lady* full frontal. According to its crew, *Yankee Lady* is the only restored B-17 flying with a properly set-up quartet of turbochargers.

Peter Garrison's homebuilt airplane *Melmoth 2*.

The modern sea kayak may be the most efficient human-powered device ever invented.
Photo credit: Pygmy Boats Co.

A Robinson 22 helicopter.
Photo credit Robinson Helicopter Company

A Hinckley Picnic Boat. The term 'picnic boat' was a nickname that stuck.
Photo credit: The Hinckley Company

The GE Evolution locomotive.
Photo Credit: John B. Carnett/*Popular Science*

The author at the helm of the Evolution, which sells for almost $2 million.
Photo Credit: John B. Carnett/*Popular Science*

The Cornwall Volunteer
ambulance the author
drives while on duty.
Photo credit: Stephan Wilkinson

Left: A Steinway piano being built in
the NYC factory. Steinway has been
turning out pianos for the last 150 years.
Below: Two nearly finished
Steinway products.
Photo credit: Steinway & Sons

Above: B-17 Thunderbird.
Photo credit: John Preacher

Left: A Vietnam-era YO-3A stealth aircraft.
Photo credit: Quiet Aircraft Association

B-17s.
Photo credit: John Preacher

The author with his falcon. The sport of falconry is 4,000 years old.
Photo credit: Susan Crandell

Sentimental Journey nose art.
Photo credit: Dave Steiner

B-17 *Yankee Lady* nose art.
Photo credit: Joe Horenkamp

Cory Bird's airplane, *Symmetry*, which took approximately 15,000 hours to design and build.
Photo credit: John B. Carnett/*Popular Science*

Three B-17s.
Photo credit: Rand Bitter

They Named It the Seahorse,
But They Called It the Dog

✿

"WHENEVER I MENTION THAT I FLEW HELICOPTERS in Vietnam," an American ex-Marine pilot said to me recently, "the first thing people say is, 'Oh, you flew the Huey.' You've got to explain to them that no, this was a different helicopter."

Boy, was it. Turns out the guy flew the Sikorsky H-34 Seahorse, known in its USMC guise as the UH-34D. D for deafening, different, drafty, dervish-slash-whirling, and just about everything but diminutive or debonair.

The Marines called it the Dog, and its pilots often referred to themselves as Dog drivers. Not because it was one, but because no American warrior would dream of calling a weapon by its government-approved name. Somewhat more irreverently, they also called it the Shuddering Shithouse. To understand the latter, you have to at least ride aboard one, preferably fly it. I've done a little of each, although I've also been told by experienced ex-Marine pilot Mike Leahy that the scatology was originally prompted by the Sikorsky CH-37, an enormous early heavy-lift helo that was powered by two Pratt & Whitney R-2800 radials. "That thing literally shook like a big dog after its bath when you approached for a landing and slowed it to a crawl," says ex-CH-37 pilot Leahy.

The UH-34D—and the CH-37—represented the end of an era. It was the era of piston-engine, clanking-parts, fling-wing, Rube Goldberg helicopter design sometimes described as "a collection of rotating and reciprocating parts all trying furiously to become random in motion." One example of the -34's archaic complexity: the single main rotorhead, about the size of a cowboy's Stetson hatbox, has what U.S. Army H-34 pilot William Walton recalls was 84 grease nipples. "And they had to be greased after every flight."

It was an era that was about to be coffin-nailed shut by humming turbine engines, sophisticated rotor systems, and then-unimaginably light materials and devices. Today, a turbine helicopter engine equivalent to the piston-engine H-34's 1,525 horsepower weighs about 25 percent of what the ironmongered UH-34D engine did. Indeed, the H-34 soldiered on for another decade as the Westland Wessex, once it was fitted with twinned Rolls-Royce Gnome turboshafts.

The H-34 was basically a bulked-up Sikorsky H-19, a late-1940s design that pioneered a unique powerplant configuration. In pre-turboshaft-engine days, a good place to find reliable, compact, high horsepower was in air-cooled radial engines. The obvious place to put such an engine would have been in the very center of a helicopter, directly under the rotor hub, low in the airframe for stability, with a vertical driveshaft. But that would have pretty much filled the cabin, making the machine pointless as a people-hauler.

Sikorsky's solution, in the H-19 and then the H-34, was to stick the radial out in the big rottweiler nose, with a driveshaft angled up and aft, passing between the flight-crew seats to the transmission and rotor hub at about a 45-degree angle. This left a boxy, unobstructed area for a cabin. Mounting the heavy engine in the nose also counterbalanced the long tailboom. With the potential of light, compact turbine engines apparent by the time the H-34 first flew, in 1954, the helo was already obsolescent and about to be obsolete.

A war, however, delayed the H-34's retirement ceremonies.

When Vietnam began to heat up, in the early 1960s, it was still a place for professionals—lifers and volunteers, many of them United States Marines. The Marine Corps has typically been at the bottom of our military-equipment food chain, making do with what the favored armed services have either cast off or haven't been able to figure out how to use. A classic albeit anomalous example was the F4U Corsair. When U.S. Navy pilots decided the Corsair's bouncy main-gear oleos made them too difficult to land on aircraft carriers, they were seconded to the Marines—who inadvertently ended up operating one of the great fighters of World War II.

So their two dozen old UH-34Ds were all that Marine Squadron HMM-362—HMM = Helicopters, Marine, Medium—had to work with when on 15 April 1962 they landed on an abandoned World War II Japanese fighter strip at Soc Trang, on the Mekong River Delta. The Ugly Angels, as they soon came to be known for their medevac missions, were eventually followed by nine more UH-34D squadrons.

South Vietnam's own air force operated a squadron of UH-34Ds handed over by the Marines, and Air America, the CIA's covert "airline" operating in Laos, also relied on the Dog. In fact, though -34s were eventually operated in combat by everybody from the Argentines to the Israelis, Air America's -34s were in combat areas longer than those of any armed forces in the world, including the USMC. "I didn't even know we were owned by the CIA," admits Air America -34 pilot Charles O. Davis. "I flew it more than 3,000 hours and never had an engine failure. By today's standards it was archaic, but as far as I'm concerned, the H-34 is the DC-3 of helicopters. When I climbed into that aircraft, it was like putting on a good, comfortable pair of shoes."

By the time the world's media had swarmed the war, in the late 1960s, the chuttering rumble of the -34's radial engine had largely been replaced by the waspy whine of the Bell UH-1 Huey. The TV nightly news—and eventually movie theaters as well, thanks in large part to *Apocalypse Now* and its Ride-of-the-Valkyries scene—resonated with the WHAP-WHAP-WHAP of the Huey's wide twin rotor blades, and most of us grew to assume that Vietnam was the Huey's war.

But in the seven years that U.S. Marine UH-34Ds were in-country, they flew as everything but gunships. They carried troops, recon teams, cargo, crates of ammunition that their crew chiefs literally kicked out the door during low passes over beleaguered landing zones, often-pointless packages, and paperwork on what were called "admin runs"—the Vietnam-helo equivalent of World War II's milk runs—chaplains ("holy helo" trips), bodies, and, perhaps most memorably, the wounded. Without the UH-34D's endless medevac shuttles, far more Americans and South Vietnamese would have died.

The Dogs were powered by the same Wright Cyclone R-1820 radial engine used in 1930s American biplane fighters, the B-17, and indeed, some DC-3s, and the aircraft had never been intended to do battle against ground troops. H-34s had no guns, no cannons, no rockets. No problem: the Marines welded up mounts for light M-60 machine guns—one on each side of the aircraft—and installed them in the field, but that was all the recoil that the airframe could take.

The H-34 had been designed to be a carrier-borne U.S. Navy antisubmarine-warfare bird, fighting a relatively neat search-and-detect sonar war at sea. Unfortunately, the aircraft's skin was made of superlight magnesium. Sikorsky had to do something to compensate for that enormous lump of engine in the nose, but magnesium did its best to become powder in the presence of salt water.

That magnesium was also to become a liability in battle. "On my second day of flying in Vietnam," recalls ex-34D pilot Seppo Hurme, "one of our -34s was shot down, and you could see it from miles away, the magnesium burned so bright. But you never had to worry about ending up a cripple. Between the avgas and the magnesium, you either walked away from a crash or you died." (As ex-HMM-363 pilot Joseph Scholle puts it, "We used to call it the world's largest flashbulb. Get a fire anywhere and drop it in the water is about all you can do.")

Nonetheless, Hurme loved the old Dog. "That big engine up front was the equivalent of a lot of armor plate and gave you more protection than there was in other helicopters. I heard of one guy who took a hit from a 57mm recoilless rifle that knocked one of the cylinders completely off. The engine kept running—rough, but they still got away. When I transitioned to Hueys, I felt naked."

Hurme had been trained to fly the Dog's replacement—the big turbine-engine, twin-rotor Boeing-Vertol CH-46—for by 1967, when Hurme arrived in Vietnam with HMM-361, the UH-34D was well on its way to an honorable retirement for the second time. But the -46s soon experienced a series of in-flight failures: they would shed their entire tail and rear-rotor pylon. As

Joseph Scholle describes it, "The H-46s would break apart right in front of the stub wings, and become a section of two H-23s." Such gallows humor is typical of helo pilots. American Newsman Harry Reasoner once wrote, "Airplane pilots are open, clear-eyed, buoyant extroverts, and helicopter pilots are brooding, introspective anticipators of trouble. They know if something bad has not happened, it is about to."

The accidents led to the CH-46's grounding, and the Marines turned back to the faithful UH-34D. Says Scholle, "Everybody wanted to go into Hueys and gunships, and when I got assigned to -34s, I thought, 'Aw, jeez— -34s, the Shuddering Shithouse.' But the part I grew to like was its reliability. We'd get more time out of our engines than the Hueys were getting. All that red-clay sand used to get sucked into their intakes and eat the turbine blades alive. We had an air cleaner, basically, like you have on a Pontiac. Take it out, bang it on the ground, rinse it in avgas, and you're back in business."

Scholle also knew of a -34's nine-cylinder Wright suddenly transitioning to an eight. "A friend of mine was doing a recon insert [dropping off a squad of Marines] way up by the Laotian border. It turned bad and he had to go back in and pick the guys right up again, and he took a lot of fire. He said the aircraft felt a little sluggish, and when they landed to let the recon team off, there was no oil on the dipstick. So the crew chief emptied the spare five-gallon can into the tank, put the cap back on, and said, Let's go. They flew back to Da Nang, and the crew chief opened the clamshell [nose] doors and said, 'Cap, you gotta see this.' One entire cylinder was missing, the piston was missing, the rod was missing—it was just a hole in the side of the block. They came all the way back from the Laotian border with it that way."

Nor were engine parts the only thing the Dog could do without. "It was one of the few helicopters that would fly with an inoperative tail rotor," says Scholle. (A helicopter's tail rotor is intended in large part to oppose the innate desire of the fuselage to rotate rapidly around, and counter to, the main rotor shaft.) "A -34 has an awful lot of side area, and as long as you're doing 45 knots, it swings around into about a 45-degree crab and stays there. It's weird, but you can fly it."

Scholle recalls, "She'd also fly without transmission fluid. Guys would have the transmission oil cooler shot out, the oil pressure went to zero, and you'd just fly it back. You do want to keep the power up, though, because once the gearbox stops, it welds itself into a single piece."

One problem with the big Wright radial was that it was easy to overspeed the engine, particularly with the rotors disengaged. Without a prop to drive, the engine in effect had no flywheel to stabilize its acceleration. "We had one copilot who hooked his lapbelt around the collective by mistake," says ex-crew chief Bobby Johns. "When he raised his seat, it pulled the collective up [and automatically added throttle]. It sounded like a double-A fuel dragster out there on the flightline. Everybody immediately knew something bad had happened." Even worse happened to a pilot that U.S. Army aviator William C. Walton remembers. "One guy in our squadron in Germany reached down to adjust his long underwear or something and hooked his arm under the collective by accident. The engine beat itself to death. Broke its mounts and fell right out onto the ramp."

For all their size, -34s were surprisingly nimble. "It was an extremely maneuverable aircraft," says one ex-Marine pilot who flew it for almost a year. "You could get into and out of landing zones where no other helos could go, but the bad news was that once there, we pilots were sitting 13 feet up in the air, and the bad guys were in a prone position, laying on the ground flat as they could or hiding behind a tree, firing back at you. You were a sitting duck in a hot LZ."

Rod Carlson remembers that. He was another rerouted CH-46 pilot, sent to HMM-361 to instead fly Dogs. Carlson drew his first night-medevac mission soon after arriving at Marble Mountain, the helicopter strip at Da Nang, flying with aircraft commander Captain Ron Sabin. Marines who were expected to die in the field before daybreak were flown out, but it was a fearsomely dangerous undertaking. Carlson and Sabin waited for a summons in the squadron ready room, where, "with the red lights on to preserve our night vision, everything was the color of clotted blood," Carlson recalls.

When the phone rang, Sabin and Carlson sprinted to their -34 and fired it up. "A constant blue-white flame from the exhaust stacks extended past my window like a huge blowtorch. Once we were airborne, Sabin flipped off the light switches overhead, and except for the flame, everything vanished in total darkness. I felt as though I were in free fall. Everything I needed to fly was gone—airspeed, altimeter, rotor and engine rpms, manifold pressure, a horizon—but the sensory inputs I needed for survival Sabin didn't seem to miss in the slightest."

Below them, Carlson says, "lights blinked like the small farms we flew over during night-training hops from Pensacola. But each was the muzzle flash of a gun being fired at us." The landing zone was hot—under fire—so Sabin told the grunts on the ground to mark its center with a small strobe.

"The standard procedure was to spiral down directly over the LZ, in order to present the smallest target for the shortest time. In daylight, this approach was dangerous. At night, I was sure it was impossible." Carlson remembers that "Sabin bottomed the collective, screwed back on the throttle, dropped the nose, and spiraled down like a duck with a shot wing. After five complete revolutions, he straightened out, and the strobe was dead ahead. I could feel him raising the nose, to slow our forward movement and twisting on full power to stop the descent."

Sabin maneuvered to put the strobe between the helicopter and the waiting Marines, but the light kept moving: the Marine carrying it had mounted it on his helmet, figuring that would make it a better beacon, and now he realized Sabin might try to land on top of him. "Ron landed with his side toward the shooting, so the exhaust stacks wouldn't be a target, and we picked up our guy. I remember as we headed back toward Marble Mountain, Sabin got on the intercom and asked the corpsman down in the cabin, 'How's he doing?' The medic said, 'I've got my hand inside his chest, but he'll make it.'"

Before his first night aloft was out, Carlson and Sabin would do it seven more times, a typical shift for a ready-when-you-are Dog.

Ron Ferrell was also a corpsman on UH-34Ds, and what he—and many a pilot—particularly appreciated about the Dog was that it had big, fat wheels and tires on long-stroke oleo struts. "We were lifting off under fire one day," Ferrell says, "and the pilot took a hit in the head just as we took off. We were nose-down, tail-up, and he had the rotors cranked up to full rpm. And then boom, we set right back down. We probably dropped a good 10 feet. I watched those struts go damn near to the ground and then spring back up. Hueys just had skids, and if that'd happened in a Huey in anything but an absolutely level attitude, you'd have pitched over.

"The copilot got control and we took off again, but it was frustrating. There's a bulkhead at the forward end of the cabin, and above that is where the pilots sit, and you can't get up there from the cabin. All I could see of the pilot was his feet, his hand hanging down, and there was blood running down the bulkhead. And I couldn't do a thing about it."

John Downing, an HMM-361 UH-34D pilot in-country, remembers that the big landing gear made it easier to get into a tight LZ. "You could stand it up and put the tailwheel on the ground, haul back on the cyclic, and get it about 40 degrees nose-high, just put the tailwheel on the ground, and it'd stop on a dime. That got me in trouble when I transitioned to the Huey, because you definitely don't want to do that in a UH-1. The first thing that hits is the tail stinger, next is the tail rotor."

Joe Scholle also loved that landing gear. "If you look at where the main-gear struts attach to the fuselage, they're basically holding up the transmission deck, because you don't want that thing coming down through the cabin. So if you land real hard and bounce, you're bouncing off the point of main structural strength in the aircraft. You try bouncing a Huey, hit tail-first, and you're probably going to nose over. But the -34, you come in nose-high, hit tailwheel first, and bounce off the main wheels, there's no damage at all. If you had to do a night medevac, it was the best way to find the ground, with the tailwheel. It was like a hen in a nest, putting her butt down on the eggs."

H-34s were the first helicopters to get a true stability-augmentation system—a kind of primitive autopilot that did its best to counter a helo's innate desire to do anything but fly straight and level. "It was called the ASE system—automatic stabilization equipment—and it gave the aircraft a little more of a stabilized feel," one pilot remembers. "But there were many occasions when it wasn't operating, so you flew without it." Well, you did if you were a sharp stick-and-rotor guy. HMM-362 door gunner and frequent unofficial copilot Bobby Johns recalls, "There were pilots who wouldn't fly it if the ASE was not engageable. It's a hands-on bird, and with the ASE working, you could set the trim and actually turn loose of the controls." Ex-Huey gunship gunner Jim Moriarty remembers that, "There were UH-34D pilots over there with a thousand hours, and we thought they were heroes. Well, now I realize 990 of those hours were with the ASE on."

With the ASE off, the aircraft was extremely sensitive to the touch. "All you had to do is think about doing something, and you've already done it," one pilot remembers. It took a fair amount of coordination to manually adjust the rpm and pitch simultaneously with a motorcycle-grip throttle on the collective while working the cyclic stick and rudder pedals at the same time. "You could overspeed it quite easily if, say, you were going into a hot LZ, taking fire," he recalls. "You had to listen to the rpm, the sound of the engine and the rotor blades, without looking at the gauges. When I went back and flew a UH-34D again a year ago, after having flown a lot of fixed-wing military and commercial aircraft in the intervening 33 years, I knew how the barnstormers felt in the 1920s, listening to the sound of the wind in the wires and feeling the slipstream on their cheeks."

The Dog that he flew is one of two still operating in a coat of flat Marine green. It is owned not by Uncle Sam but by that very same Huey gunner, James Moriarty, now a wealthy Houston, Texas, lawyer who loves the Corps enough to have spent an unconscionable amount of his own money restoring the 40-year-old Dog so that he can operate it as a living, breathing, shuddering, fluttering, flying Marine memorial. ("I sue big companies that cheat people," he says with in-your-face pride. "Erin Brockovich is my hero.")

Moriarty takes his YL-42—the call letters of a fatally crashed HMM-362 Ugly Angels bird—to air shows all over the United States. It hasn't been restored to compulsive museum standards, but renovated to the condition of a hardworking UH-34D just as it might have looked parked on the ramp at Soc Trang or Marble Mountain waiting to insert a recon team, fly a medevac, or resupply a squad in desperate need of ammo. The cabin is cluttered with toolboxes and spares, the slightly askew clamshell nose doors are safe-tied closed with a bungee cord just as was often done in Vietnam, there's a small puddle of red hydraulic fluid on the cabin floor, and even an M-60—deactivated, of course— on its swivel mount in the main door.

Ex-corpsman Ron Ferrell wonders what it would be like to ride a -34 again, 35 years later. "If I could get back into one, I'd do it in a heartbeat, if only to hear the sound and smell the exhaust. But it still wouldn't be quite right. It's like watching a movie. Even if it's realistic, you see it and hear it . . . and that's it. You can't smell it, you can't feel the heat, the exhaust, the concussion of the explosions, hear real people who are really hurt, smell the machine, smell the cordite, the vegetation, the blood sera. You can't get that by just going out to the airport and getting a ride. I'd love it, but it wouldn't be the same."

Moriarty exhibits YL-42 not because he's trying to re-create the Vietnam war Confederate Air Force-fashion, or is particularly an aviation buff, or has much truck with the warbirders who restore pristine mine's-bigger-than-yours playthings. He does it because, "We lost 58,000 people over there, and all of their mothers and fathers, sons and daughters, brothers and sisters, wives and husbands still think about them all the time, and it's important that they never be forgotten. I survived Vietnam, I thrived, I got help with my college education, and my country doesn't owe me a thing. But those families still wonder, was it in vain, was it worth it, did it have a meaning?"

Moriarty served three combat tours in Vietnam as a Huey gunner and ultimately became a sergeant. (He now laughs that he's the only sergeant pilot in the Marines.) "We used to occasionally see the UH-34s at Marble Mountain, but I thought those old radial engines were totally obsolete. Hell, I was flying in turbines. The guys still flying -34s were fighting a different war than we were. I'm sitting up there in a gunship with eight machine guns and 38 2.75-inch rockets just dyin' for some sonofabitch to take a shot at us, and they were down there on the ground, in the weeds, the pilots sitting way up in the air framed in a big window with everybody shooting at them."

Jim claims to be the only Huey gunner ever to shoot down an aircraft in Vietnam. Unfortunately, it was his own Huey. Fortunately, he exaggerates, for he only almost shot it down. Burning out the barrels of his M-60s during one air/ground tussle, he never noticed that he'd snapped off one of the gun's traverse stops. His own bullets then shrapneled part of the mount, and the pieces honeycombed the tailcone, severing four of the seven strands of the cable that controlled the tail rotor.

Moriarty's claim might be contested by door gunner Mike Leahy, who recalls one UH-34D flight in which the action got so hot that he managed to shoot through his own bird's main rotor blades. "The mechs back at the maintenance facility recorded the bullet holes as 'combat damage,'" he remembers.

Leahy served as a gunner even though he was a major and a rated helicopter pilot. As a watercolorist and the executive officer of the U.S. Marines' combat-art program, he decided that the best vantage point from which to view scenes he could later paint was from the door of a Dog—but that meant he had to work, not just ride. "As a fully functioning crew member of a UH-34D, I also sometimes had to load KIAs into our chopper for the doleful trip back to Da Nang," Leahy recalls. "We once were so full that I had to hold the upper part of a body bag in my lap, as I manned my M-60 on takeoff, and I could feel the trooper's still-warm torso. It was one of the most moving situations I ever went through."

George Twardzik was an HMM-163 "Angry Eyes" door gunner, a squadron so named for the glaring samurai eyeballs painted on their UH-34D's clamshell nose doors. Twardzik remembers the day in March of 1966 when "we received a frantic call from an Army Special Forces unit of about 220 who were under siege in the A Shau Valley from an enemy force estimated at 3,000 men. An Army helo sped in to assist them and was promptly shot down. It was decided the risks were too great and that all units should stay away from the A Shau outpost. For three days, we could hear the troopers begging over the radio for medevacs, ammo, and water."

Finally, Twardzik's squadron skipper could take no more. He strode to his -34, turned to the anxiously watching pilots and aircrew, and announced that he was going for a ride, and if anyone wanted to join him, he wouldn't stop them.

The entire squadron fired up and headed for A Shau. Unfortunately, the skipper and three other -34s in the first wave were immediately blown out of the sky. Since it was too late in the day to do any more, the surviving -34s returned to Phu Bai to regroup. First thing the next morning, the Angry Eyes returned to the LZ like angry hornets and began pulling out soldiers.

Twardzik remembers that during one extract, his aircraft was taking fire from a .50-caliber machine gun, and eventually, it found them. Twardzik took a ricochet squarely on his flak jacket, and during liftoff, the impact blew him out of the door and into space. "My gunner's safety belt was hooked to a D-ring on the deck, and when it reached the end of its travel, it snapped me right back into the cabin."

Another round ignited the ever-present five-gallon can of spare engine oil. The pilot autorotated down into a clearing, and they pitched the flaming can out and extinguished the remains of the fire. With the engine restarted and the rotors re-engaged, "we took off, dragging the main gear through the trees as we headed back to Phu Bai. I got out of the -34 to view the damage, and the aircraft was literally sieved with bullet holes." The Angry Eyes nonetheless managed to save every one of the HMM-163 aircrew who'd gone down the day before, as well as 190 of the 220 Special Forces soldiers. "The LZ looked like an aircraft parking lot by the time we were finished," Twardzik says, "and when it was over, HMM-163 had only five flyable helos left on the flightline."

Jim Moriarty's own -34 was originally manufactured for the U.S. Navy, as an HSS-1N ASW bird. He found it corroding in a New England farm field and bought it without realizing what he was in for, but intent on bringing back to life a simulacrum of a real Vietnam-era aircraft. "To say I knew absolutely nothing about the business of aircraft restoration is being too kind. I paid $45,000 for it, figured we'd fill it with gas and fly away. What did I know?"

Certainly Moriarty himself wouldn't fly it away, since he had never flown a helicopter, but in any event, the hulk was trucked to a restoration shop in Tucson, Arizona, and the job began. "I had no reason to believe that it was anything I could ever fly," he admits. "This is one huge, powerful, noisy, intimidating machine." But soon, Moriarty added a rotary-wing rating to his pilot's license, learning to fly helos in a little two-seat Hiller, "and at the end of my first trip riding in the left [helicopter copilot] seat, I began to figure maybe I could learn to fly this thing. If you can fly an underpowered little Hiller, you can do aerobatics with an H-34, it's so powerful."

He wouldn't have been the first pilot to do aerobatics in a -34. "Oh yeah, they were maneuverable," laughs Joe Scholle. "I remember a guy did a couple of rolls and then looped it, for the benefit of the A4 and F4 pilots sitting on the beach at Chu Lai, in late '67. Of course, you don't get a real circle out of it, it looks more like a backward nine. But the -34 was pretty forgiving."

Forgiving enough to allow even me, a fixed-wing-only pilot, to take cyclic in hand and learn, very briefly, just how hard it is to fly a helicopter.

It was, for Houston, a frigid day—high 40s F.—when I showed up to go for a ride with Moriarty. "We've got 120-grade oil in it, and it was about the consistency of lard when I got out here this morning," Jim complained. "Took me $500 worth of preheat to get it to this point." The Wright groaned through a few turns, and one cylinder fired as Moriarty flailed from starter to mag switch to primer to throttle. Again, a single pop. Again. Finally, the engine broke into the classic radial-engine start-up lope. CHUFF-chuff-chuff . . . clatter clatter . . . is it gonna die? . . . CHUFF-chuff . . . clatter clatter . . . chuff . . . pop fart pop pop pop . . . KATUMP KATUMP KATUMPKATUMPKATUMP . . . Nope, it's gonna run.

Moriarty is quick to deprecate the fact that he's a rich lawyer and pilot ("meaning I'm above real work," he laughs), but he takes quite seriously the task of operating "a 40-year-old aircraft with parts just looking for an opportunity to break." He prepared for flying the Dog in part by attending a week-long North American T-28 Trojan ground school, since T-28s use the same engine the UH-34D does, and it was also a good opportunity to check out the warbird world.

"One of the things my instructor told me was that if you plan to fly a complex, old aircraft safely, you'd better be willing to get your hands dirty and learn to work on it yourself. I thought this was bullshit. I can afford all the mechanics that I want. But he was right and I was wrong. If you want to fly a

40-year-old aircraft, you need to know it backward and forward, you have to be willing to get your hands dirty, and you should not think for an instant that run-of-the-mill mechanics have any idea how willing to fall out of the sky these old aircraft are."

Big talk? No. On the first leg of our day's flying, we got a bright-red engine chip light just as YL-42 shuddered into a hover to touch down at our first en-route stop. In less time than it takes to negotiate the drive-thru at a Houston McDonald's, Moriarty had the clamshells open, a canvas toolbag spread beside him, the safety wire cut, and the chip detector unbolted from the crankcase. Followed by two quarts of hot, black oil. Moriarty may not be the only big-time lawyer in Houston with dirty fingernails, but he's probably the only one to earn them lying on his back under a Wright Cyclone.

He could as easily have summoned a mechanic—we had landed at a busy general-aviation airport just outside Houston—but Moriarty preferred to do it himself. "Nobody has to fly with me who doesn't want to," he says, "and I can get you a ride back to Houston in minutes, but this is all we've got." He points out a single 3mm-long hint of metal the thickness of a strand of fine steel wool that has bridged the detector's electrified gap. It is almost certainly a stray sliver from the valve job the engine has just undergone, and I can't help but think of the little Porsche 911 racecar engine I have just rebuilt. After running the car for 30 minutes on jackstands and then cutting the oil filter open, I found enough post-overhaul trash to cover a fingernail.

Jim and JT Nelson, his crew chief, indulge in yet more helicopter gallows humor. "Aw, ya really don't have to worry about chips until they're big enough to read a part number off'n 'em," JT laughs. There must be a reason why fixed-wing pilots don't make such jokes about their machinery, but you can't be around a fling-wing pilot very long without hearing that "Helicopters don't actually fly, they just beat the air into submission." Or that "Helicopters are so ugly, the ground simply repels them." Or that "If something hasn't broken on a helicopter, it simply means it's about to." Or, "Never fly in anything that has wings traveling faster than the fuselage."

I chose to fly rather than drive, even though the wings were obviously whirling about my head as I glanced up through the cockpit Plexiglas. It was disquieting to see that huge, complex mechanism spinning around at a rate that seemed at the same time too slow to hold us aloft yet quite fast enough to be dangerous. It was the equivalent of somehow being able to watch an airplane's main spar working, bending, twisting, wearing . . . which, of course, we fixed-wing folk never get the opportunity to do, thank you very much.

"Rwaaarrwaaar, waarwar. Rrrrwarr," Moriarty yelled into the headset intercom. Between my aviation-heightened deafness and the hysteria of a Cyclone-engine helo cockpit, it was all I could do to figure out that he was saying, "Take the stick. Try it."

After a decade of flying a Falco, I had finally learned some fingertip control delicacy, though it took awhile to overcome the brutishness instilled by a lifetime of Cessnas, Pipers, Beeches, and Mooneys—all of them airplanes that require the finesse of an NFL center. But would that apply to a 1,525-hp, 13,300-pounds-gross cargo aircraft? Well, I took the stick.

Mistake.

The world assumes that the pyramid of aviation proficiency rises from a broad base of ordinary lightplane pilots up through bigger and faster airplanes to a pointy summit upon which sit . . . well, people who fly pointy-nose airplanes. Jets. 747s. Concordes. Tom Cruise Tomcats. F-15s and Tornados. They're wrong. The most tactilely skilled pilots in the world, in terms of hand/eye coordination and the physical touch to guide a heavier-than-air machine through the sky, are helicopter pilots.

I quickly learned that, through the simple act of trying to keep the aircraft straight and level. Helicopters want to diverge. They are by tradition and temperament unstable. Push a trimmed airplane's stick and it resists, insisting upon returning you to level flight. Push a helo's cyclic and the damn thing is hell-bent to bunt. Pull it back a bit, as you of course do, and you are immediately in a world of pilot-induced oscillations. If a well-trimmed airplane is equivalent to a train running on tracks, a helicopter is a unicycle. And an unstable old 1950s piston-engine helo is like riding a unicycle while balancing a broomstick on your chin. With a kitchen table atop it.

"With all your hours and ratings and fixed-wing experience," Moriarty later said, "if I'd had a heart attack up there, you'd have had about 40 seconds to live." He's right. At best, I'd have been able—with great, constant effort—to maintain 500 feet and an eastbound heading until we ran out of gas.

Today, Moriarty has logged over 400 hours in his YL-42. He and JT fly in full Marine flight suits, complete with flight-crew wings and HMM-362 squadron patches and insignia. Moriarty also straps on a loaded survival vest before clambering way up into the cockpit. Uh oh, middle-aged men playing boy soldiers, I initially think, but in fact it's because of Moriarty's respect for the tradition, history, and sacrifice that his YL-42 represents. "One of the rules of the aircraft," he says, "is that when you fly or crew it, you wear the uniform. Not because I think it's fun but because I want to honor the people who flew them. I will not fly this aircraft in shorts or jeans or tennis shoes."

"Go to air shows and you'll see a good representation of World War II and modern-day aircraft, but the missing link seems to be Vietnam-era aircraft, especially helicopters," says Roger Herman. "So Jim's aircraft really attracts a crowd. He spent all his own money to buy and restore it, and it was quite a chunk of change. And he's dedicated to promoting the history of the Marine Corps and getting people to see things just the way they used to be. At a lot of the shows, you'll see aircraft owners charging money to let people into the cockpit or whatever, and that's something Jim absolutely won't do. I think that's great."

So does Moriarty. He guesses that last year, 20,000 people clambered through and around his YL-42 at various U.S. air shows. "At first, I wondered, should I put a rope around it, only let certain people get close to it?" he admits. "But I decided no, that wasn't going to be its mission. You see kids up in the cockpit, their feet can't even reach the floor, and you can tell they imagine themselves as people someday willing to fight for their country, as people who want to care for and protect others. They need to be able to touch that dream. I want people to crawl all over that aircraft; I want them to know that brave men flew them. And that this aircraft still flies."

It sure does. Semper Fi, Sarge.

An afterword:

Unfortunately, Jim Moriarty's handsome helo came to grief in the spring of 2002, when Moriarty taxied six inches too close to a light stanchion at a small airport in Mississippi. When the huge rotor hit the pole, Moriarty recalls, "We started shedding parts, some of which flew 500 yards. It felt like a combination of riding a bronc and getting rear-ended. Parts went everywhere. After everything came to a gradual stop, I shut down the engine, and verified that no one had been injured."

The damage, however, was far too expensive to economically repair, for even the helicopter's airframe had been ruined by the extreme vibrations. Moriarty donated what remained of YL-42 to a group restoring another ex-Marine UH-34D and immediately set to work on his next project: making flyable a two-seat McDonnell Douglas TA-4 Skyhawk attack jet, which today is the only airworthy TA-4 in the country. "Naturally, it is painted in USMC colors," Moriarty proudly points out. ✿

PICNIC BOAT

❋

MANY MODERN POWERBOATS, TO MY LANDLUBBERLY EYE, have all the style and grace of a Motel Six plastic bath/shower unit, with which they have much in common in terms of construction. A classy, super-expensive exception is a motorboat made by the small Maine company, Hinckley. It is a 36-footer simply known, in a world of boats named after killer fish, as the Picnic Boat.

Nobody intended for it to be called that. It was Hinckley's original nickname for the prototype, since they figured it would be a day boat that people would use mainly to go on picnics. "Nobody expected the product to grow so much," admits Hinckley's vice president of sales, Edward Roberts.

The Picnic Boat has become the ultimate rich boater's toy, in part because of the tastefulness of its traditional Maine lobster-boat lines, in part because of some surprisingly high technology hidden within it, and not incidentally, because of its astronomic price—$480,000, and up to $520,000 with all the options and custom details you can buy—for a day boat that sleeps two. That makes it, at $1,180 per foot of length, the most expensive semiproduction pleasure boat you can buy.

But that's a bargain. It takes 60 man-hours for Porsche to assemble a $90,000 911 coupe (admittedly from supplier-provided parts). It takes 8,000 man-hours to build a Picnic Boat. If Hinckley were charging Porsche rates, they'd ask $12 million for a Picnic Boat.

Either way, that's for a market of buyers who don't intend to spend Saturday afternoons at the boatyard scraping barnacles and drinking beer. So for

$16,000 a year, typically, Hinckley will come pick up your Picnic Boat when you're done for the summer, truck or sail it to the nearest Hinckley boatyard, service all of the boat's systems, and store it for the winter. Expensive? Not really. It's about what you'd pay for the FAA-mandated annual inspection of a twin-engine private plane (which, I suspect, a number of Picnic Boat buyers also own).

"It's like owning a fine collectible car," Ed Roberts says. "You don't take it out in the rain or snow, and you're going to be very meticulous about having it taken care of." After all, you don't take your Ferrari to the we-fix-flats shop.

Lots of powerboats are at best floating Corvettes and Vipers, at worst plastic-clad Pontiacs of the sea. But if the Picnic Boat were a car, it would be a 1961 Ferrari 250GT Berlinetta Lusso. Both have the same powerful but re-strained swoop from stubby tail past clean flanks to a rising, massive, pow-erful nose. And both are shapes that were once so right yet today look, at least to some of us, appealingly archaic. It fascinates me that the perforated beige headliner in a Picnic Boat's cabin is identical to that of a 40-year-old Porsche 911.

Hinckleys are of course built entirely by hand. They appear to be made of wood, for the hulls are usually painted a traditional dark yacht blue, and the perfect gloss—it's the same durable, expensive Awl-Grip enamel used on airliners—is far shinier and more reflective than gel-coated fiberglass. The fact that there's lots of teak brightwork showing, and that Hinckley's var-nishing technique is something of a company secret, also abets that wooden-boat impression.

But a Picnic Boat's hull is actually a rugged sandwich of Kevlar, balsa, PVC, and carbon fiber, manufactured in a process called SCRIMP (Seeman's Composite Resin Infusion Molding Process).

The most basic way to build something of fiberglass is to spread layers of fiberglass into a female mold, wet them down with epoxy resin, and let the assemblage cure. When it's dry, you pop it out of the mold. It's durable but heavy, because lots of extra resin was needed to make sure all the glass weaves were fully impregnated.

Or you can lay the glass—and, typically, carbon-fiber reinforcement if you go to this much trouble—into the mold wet and then cover it with plastic sheeting, seal the edges, and pump the air out, squeezing a lot of the excess resin out of the cloth and to the edges. Called vacuum-bagging, it's better, but not best.

SCRIMP means you put everything down dry—Kevlar, carbon fiber, fiberglass, core—and "bag" it dry, then pull a vacuum and inject an exact quantity of resin at various points into the bag. The vacuum sucks the resin thoroughly throughout the layup, and you end up with a strong and light composite product.

An ancillary benefit is that also trapped in the bag is much of the glue-sniffer aroma that fills a less-sophisticated boat shop as workers stagger around with melting brains. Parts of the Hinckley factory are scarily odiforous, to be sure, but nothing like what you'll find in the warehouse factories that dot the back streets of Fort Lauderdale.

But the Picnic Boat's most impressive technology is controlled by a big teak-knobbed, rubber-booted joystick just to the right of the captain's articulated chair—a Stidd seat originally designed for long-range military patrol boats where coxswain comfort is important—upholstered in white bizjet leather soft as Jell-O. Each of the seats costs Hinckley $8,000. Though the Picnic Boat might look like it should go *tump-tump-tump* as it putts along at six knots from lobster pot to lobster pot, it in fact has a fully planing bottom, a 440-hp Yanmar straight-six diesel big as a refrigerator, and a max speed of 34 mph via a pivoting New Zealand-built Hamilton jet drive controlled largely by that joystick.

New Zealanders are today the world's jet-drive go-to guys, but the water-jet concept was developed in Ohio, just before World War II, by a company that built launch-size fireboats. They figured that they already had a high-pressure pump aboard to spray water on fires, so why not also use it to shoot water out the back of the boat to get to the fire? Worked just fine, but after making a run of small jet-drive fireboats for the Coast Guard during the war, the concept lay dormant until the mid-1990s, when it reappeared largely in big commercial craft and little jet-skis. Hinckley was the pioneer in putting it aboard midsize pleasure boats.

"The first Picnic Boat we built was a conventional prop-driven boat," says Robert Hinckley, whose father founded Hinckley Yachts. "When we put the jet version in the water and started fooling around with it, we went, Whoa! This is the way to go."

Hinckley's controller, called a JetStick, works through a dedicated Rolls-Royce Marine computer—they also make electronics for Aegis-class missile cruisers—and controls the forward-and-reverse and side-to-side direction-controlling thrust of the jet at the base of the boat's transom, by shifting the "bucket" that diverts the high-pressure water stream to the front or rear and by moving the jet itself from side to side like a conventional boat's rudder.

In docking mode, it coordinates with all of that the effect of the Picnic Boat's bow thruster, so the boat can be moved absolutely sideways. Think of the Picnic Boat as a car that can pull up next to a parallel-parking space, turn all four wheels 90 degrees, and slide in laterally.

"What makes the system special is that it doesn't make the boat idiot-proof, but it does make it foolproof," says Hinckley electrical engineer Fred Berry. A propeller-driven boat is challenging to maneuver because you need to have water flowing over the rudder to have directional control, which generally means the boat must be moving through the water. Twin screws give you the capability to maneuver a certain amount with asymmetric power, but docking a single-screw boat is equivalent to landing an airplane: it's hard to stop in the middle. You have to plan ahead and establish a "glide path" all the way to where you want to end up dockside, taking into account wind and current.

There's a substantial menu of JetStick capabilities. In docking mode, if you get confused—and many Picnic Boat buyers are wealthy novices—all you need to do is let go of the JetStick and the boat reverts to an instant "hover" setting, dead in the water with the power automatically balanced front and rear, no matter where you've set the throttle (a separate control).

Funny thing: Hinckley was once a company that made superb sail and motor yachts for serious boaters, but the business has changed, and the Picnic

Boat is part of it. "Twenty-five years ago, if you were a boater, boating was your life," mused sales VP Ed Roberts. "When you came home from work on Friday, your wife had the boat stocked and ready to go for the weekend. Your Christmas presents were all boat-related. Boating was what you did. Today, the high-end powerboat is just one of many things vying for a person's recreational time."

Meaning that a number of them are the kind of boater who puts total trust in a Picnic Boat's big, totally automated, moving-map GPS display, who uses a laminated not-for-navigation placemat copped from a Nantucket restaurant as their only chart, and who may not have the faintest idea how to dock a boat that won't hover and go sideways.

"We have had two Picnic Boats go over the breakwater at Nantucket at high tide," Bob Hinckley admits. "The boat has only 18 inches of draft, so they got a little casual, but there's only about six inches of water at that point. They bounced over, but the boat's pretty forgiving. With a traditional boat, you'd have taken all the running gear off, but they used the boats until their vacations were over, then we fixed 'em."

But hey, there's a new generation of water rats coming along. A Hinckley employee confided that they've sold Picnic Boats to couples whose 12-year-olds drive them better than they do, since to the kids, the JetStick makes it just a living video game. All they'll need is blasters to get rid of those pesky submerged breakwaters. ✿

SYMMETRY

✳

WHEN A TINY, YELLOW, BARELY-TWO-SEATER called *Symmetry* flew for the first time in April of 2003, a machine that is arguably the most finely crafted, handmade human artifact of its size took to the air after a 14-year design-and-build process. Certainly I'd wager that *Symmetry* comes closer to perfection than any other homebuilt airplane in the world.

I've seen a lot of homebuilts and made one myself—an Italian-designed Falco that some people, stroking my ego, said was a pretty obsessively constructed beauty. Parked next to *Symmetry*, however, my spruce speedster would have looked like a sorry piece of plywood trash. And my "obsessiveness" would have seemed like slapdash carpentry compared to builder Cory Bird's pursuit of perfection.

Quite simply, *Symmetry* is not only symmetrical but flat flawless.

The airplane's fuselage is as lithe and sinuous as a tango dancer's outthrust leg. The design began with a 1:10 scale model that Bird sculpted from Styrofoam. He then took a single measurement—the width of his shoulders—and worked on out from there. Not a single computer was used anywhere during the design process, which was entirely intuitive. As Bird's wife Patti puts it, "I once asked Cory why he made the airplane that shape, and he said, 'I envisioned myself as an air molecule and wondered where I'd want to go.'" And where an air molecule most comfortably flowed became *Symmetry*'s shape.

Its wings are literally as straight and true as a draftsman's steel rule, which to anybody who has built an airplane is probably Bird's single most awesome accomplishment. Sight down a typical trailing edge from tip to root, homebuilt airplane or factory job, and the inevitable variations become hugely apparent.

Symmetry's nine coats of hand-rubbed polyurethane enamel (seven pigmented, two clear) make the $60,000 paint jobs on many a concours-winning collector car look like Earl Scheib re-sprays. Under harsh fluorescent hangar lights that would make even a brand-new Mercedes appear to have been painted with a broom, *Symmetry* reveals nary a ripple nor a flaw.

When I first saw *Symmetry*, in its hangar at Mojave Airport, in California's high desert, I peered into the uncowled engine compartment and marveled at the amount of extra machinery that extended back into the space between the four-cylinder, 200-hp Lycoming engine and what I assumed was the front of the cockpit—hoses, tubing, pumps, wires, an extra pair of magnetos. . . .

Wait a minute; extra magnetos? Then I realized that *Symmetry*'s stainless-steel firewall had been polished to an absolute mirror finish and was simply reflecting the back of the engine, fooling my eye into seeing space where in fact there was hard steel.

All aircraft are pockmarked with access panels, inspection plates, fairings, fillets, and fasteners. Except for this one. Half a dozen tiny screwheads and a few hairline seams are the only interruptions in *Symmetry*'s glassy surface. The pitot tube is a minute opening in the nose of the left wheelpant. The fuel-tank drains on the belly are what the industry calls "flush" drains, with only an eighth-inch protrusion beyond the skin of the airplane. An eighth-inch too far for Cory Bird, who totally re-machined the stock push-up drains and mounted them in bushings that are, well, flush, man. Even the necessary cowling access door for the oil dipstick has no visible fastener but is opened by springing an internal latch that requires reaching into a cowling air intake.

And that entire cowling is itself held in place not by dozens of unsightly machine screws, as is typical, but solely by a circumferential tongue that fits so perfectly into a groove around the firewall that the two cowling halves need to be popped free with a slap to release the air pressure when they are removed.

The airplane's few seams—where the cowling and canopy mate with the fuselage, mainly—are so narrow, true, and temperature-sensitive that at 70 degrees, they literally disappear, becoming airtight and invisible. Even the gaps between the control surfaces and the wing and tail are so narrow that a thick business card barely fits. A shirt cardboard wouldn't make it.

Yet much of what is remarkable about *Symmetry* can't even be seen, normally. The hidden backside of every piece of the airplane is either perfectly painted, polished, plated, or carbon-fibered. Nothing is left raw and unfinished. The four-cylinder, 200-hp Lycoming engine is not only tightly cowled but also shrouded by an artful carbon-fiber downdraft-cooling airbox. The weave of its strengthening layer of cloth is as perfectly aligned as the herringbones on a Savile Row suit. Not that it matters, since the cloth is just as strong no matter which way the pattern points, but Bird would have nothing less than visual perfection as well.

The airbox is so well-designed that even though its cooling inlets are tiny—for less drag—the hot air exhausts out the bottom so efficiently that no messy cowl flaps are needed. (Most high-performance airplanes have these small, door-like appendages to allow for greater airflow when they're manually opened during slow-speed, high-power situations such as takeoff and climb.)

Cory Bird is an aeronautical engineer. Don't ask to see his degree, for he doesn't have one. Doesn't need one. He works as a manufacturing design engineer for Burt Rutan, the iconoclastic aircraft designer and builder who generally won't even talk to fabricators and engineers from "the aerospace industry." Those programmed professionals have already been poisoned, Rutan feels, by the rules and regulations that dictate how airplanes are built. Lots of rivets in nice straight lines, wings in front and tail in back, tried and true, tested and by the book.

Symmetry is Bird's résumé. "I built it in part to enhance my career at [Rutan's company,] Scaled Composites," he says. "I used it to demonstrate my ability to do every step of the design, engineering, and fabrication of an airplane. My boss now knows that I can do all that.

"Also, working at Scaled, you're not supposed to ask a fabricator to do something that you can't do yourself. One of my motives in building a flawless airplane was to gain respect from the shop guys, so they'll listen to me carefully next time I ask them to do something. You can cut a line absolutely, perfectly straight or you can cut it crooked. It takes the same length of time to cut it either way. But doing it perfectly saves all the time you'd spend correcting the bad cut. I wanted to demonstrate that perfection doesn't take any longer."

Still, it did take Cory Bird what he estimates is about 15,000 hours to design and build *Symmetry*. It cost him $40,000 in materials for the airframe and instruments, another $20,000 for the overhauled engine. (Had Cory included in the ledger an hourly fee for 15,000 hours of work by a skilled aerospace engineer and fabricator, the bill for Bird's bird would have been approximately $1,500,000.)

"Is it perfect?" Cory muses. "It was perfect when we unveiled it, before it flew. Perfect. Since then, I've gotten a few dings and chips, overheated the brakes, and softened the weave in the wheel fairings, and I know that's going to happen. Someday, I'll get it flying as fast as it's going to go, make all the mods needed to accomplish that, and maybe then I'll take it all apart, restore it, and make it perfect again."

Is Bird a crackpot compulsive? Does he have the world's neatest sock drawer? Wife Patti laughs. "His hangar, his airplane, and his boat are perfect, but he's totally sloppy around the house. It's only the things that are important to him that bring out this trait."

"That's true," Cory agrees. "You focus on what's important and let everything else slide. It's why Burt Rutan wears only blue denim shirts—so he doesn't have to waste time thinking about what to put on in the morning. It's why we live way out here in Mojave, in the middle of the desert—so I'm only four minutes from the airplane and don't have to waste time traveling."

Patti travels right behind Cory, like a motorcycle rider on a pillion seat, her legs wrapped around his hips. The seat has been precisely molded to her shape, and with Patti aboard at her present lissome weight, the airplane is at the absolute aft limit of its safe weight-and-balance envelope. (What better excuse to work out faithfully and eat intelligently?)

In a culture that disproportionately rewards any number of evanescent qualities—large breasts, the ability to throw a ball through a hoop, being hugely amusing in front of a TV camera—the Cory Birds of the world are true treasures, reminding us that perseverance, ingenuity, and workmanship can produce treasures that will truly endure. ✸

KAYAKS

✦

THERE ARE PLANERS, AND THERE ARE DISPLACERS. Among waterpeople, the planers are the ones whose boats are not of the water but upon it, bouncing along the surface like flung flat rocks. In the most extreme examples—hydroplane raceboats—they are mated to the water only by the lower half of the spinning prop's disc and the occasional touch of a sponson. Very fast. Very raucous. Very horsepower-intensive.

I'm a displacer. The hulls I fancy displace their weight in water. They float.

But the hulls I really like are mere splinters, boats of extreme fineness— fineness being the ratio between length and beam, the higher the better. A floating splinter is the machine a man should power when he's on the water. When I was in college, I helped row a shell, a spectacularly fine boat, as part of an eight-oared crew. At least I did until I proved too puny even for the bow position on the lightweight team, which was where they put the smallest kid who was strong enough that he helped to keep the boat's average per-oarsman weight down to 150 pounds. (That crew went on to win the world lightweight championship without me, I must admit.)

So I switched to "singles"—sculls—little waterbugs so tiny and narrow that you sat atop them on a sliding seat, feet laced to footrests. The only thing that keeps them upright is the blades of the two oars in the water like outriggers. They are fast, silent, efficient, and as tippy as a Suzuki Samurai.

That was then. Now I paddle kayaks, those funny boats that most people think are the word-association answer to "Eskimo." I'd always thought kayaks the purview not of Inuits but of men and women who probably were also birdwatchers or nudists—unsmiling folk who take the outdoors seriously. But today, kayaking is one of the fastest-growing sports in the country, as people with a sense of humor as well as adventure discover what splendid little machines these boats are.

The modern sea kayak is one of the most efficient human-powered devices ever invented, if you include in the measure of efficiency not only simplicity and speed but utility. They are not only beautiful and fast—they're useful transportation. Bicycles are far faster than kayaks, but they're also a lot more complex. There are faster oared boats, even—those sculls of my distant youth—but they are flat-water craft of absolutely no utility. And though a competition sailplane is the fastest human-"powered" machine of all, it's worthless unless the sun is available to create thermals.

Kayaks have crossed oceans—something no bicycle or sailplane has yet accomplished—and can do an easy four or five mph all day (or night) long, will carry a substantial amount of gear in watertight compartments fore and aft, and shelter their paddler as well as a wetsuit would. You literally plug yourself into a kayak and then lock yourself in with a "spray skirt" that you wear like a too-short suspendered dress, fastened tight around your chest, that then attaches to the teardrop-shaped cockpit coaming and forms a water-resistant seal. (A kayaker on dry land, skirt dangling long in front and short in the rear, looks as silly as a kilted Scot unaware that his buns are out in the breeze.)

The faster the kayak, the less stable it is: the higher the fineness ratio, the tippier the boat. During a group paddle off La Paz, in Baja California, I once found myself in a snarky 19-foot racing single that nobody else wanted, a slippery fiberglass toothpick that spit me out three times before I made partial peace with it. A skilled kayaker could have done far better by "bracing" with the paddle, slapping the water to stop the roll, but I was about as stable as a first-time unicycle rider.

Stable vehicles are no fun. Fat SUVs are stable. A 747 is as stable as Uranus. A modern cruise ship is artificially stabilized on every axis to help the bluehairs keep fox-trotting. Newsboy Schwinns are vastly more stable than titanium Serrotas.

If you're a gunfighter, you want an F-16, so unstable that it has to be flown by a computer. If you want to enjoy driving, you need a car that can be throttle-steered and rotated, not just aimed. And to me, part of the fun of kayaking is balancing the boat. For a while, I was like a novice pilot, watching the horizon to stay level. "Lookatthedolphins!" my Baja floatmates would yell, but I didn't dare. I had what kayakers call "the shakes," my boat wiggling under me as fast as I could make my butt countercorrect.

If a skilled kayaker tips over, usually as a result of heavy seas, he or she will casually do an Eskimo roll—continuing the capsize through 360 degrees and popping back out of the water—but when I went over, I was left to scramble my way to the surface after hanging from the inverted hull, tied to the boat by my spray skirt. It releases easily enough, with the pull of a toggle, but upside-down underwater is not a nice place to be. Particularly if you forget to first take a deep breath.

Nanook of the North would be shocked to see, thump, and smell a modern kayak. Despite its arcane sealskin-and-whalebone construction traditions, the sport has become as high-tech as rock climbing or mountain biking. Though there are still some wooden kayaks—particularly those available as do-it-yourself kits for homebuilders—most are today made of Kevlar, carbon fiber, fiberglass, and for all I know, titanium.

Paddle design and construction might seem as simple as making rowboat oars, but there's both art and craft to shaping what is literally a kayak's propeller. Some of the lightest and most expensive (and most delicate) are also made from carbon fiber or graphite, and the twist, shape, and blade area can make as much difference as they do in an airplane's prop. And you'd be amazed by how much ounces can matter when you're swinging the thing down, up, and around all day long.

A well-dressed paddler can easily be toting a couple of thousand dollars' worth of gear before even getting into the boat—paddle, spray skirt, state-of-the-art kayaking life vest, emergency strobe, GPS, waterproof PDA (for everything from consulting tide-table programs to playing games when you get marooned), and other accessories.

Knife-like and narrow as a sea kayak's hull is, there's still ample opportunity to equip it with a lightweight, foot-operated bilge pump, deck-mounted maritime compass, retractable skeg or rudder, padded and cushioned special seat with thigh braces and hip padding to lock you firmly into the machine, and a wide variety of must-haves bungeed to the deck, fore and aft.

Thus accoutered, you're the master of a vessel with the beauty, purity, purposefulness, and shape of a fine broadsword. Just be sure to sit up straight. ✿

How Pilots Think

✦

Until Al-Qaedas answered learn-to-fly ads and went shopping for cropdusters, most Americans assumed that private flying was a hobby pursued by rich people in "Piper Cubs." But media scrutiny brought the aviation experts out of the woodwork. Like the lady who wrote to *The New York Times* after a 15-year-old slam-dunked a Cessna into some Tampa plate glass to suggest that all U.S. flight schools be closed and people forget about learning to fly unless they joined the Air Force.

Unfortunately, a competent, talented, and intelligent group of enthusiasts was maligned by this rush to judgment. General aviation, which is what they call flying that's neither military nor airline, is largely inhabited by people who have labored mightily—certainly in comparison to what it takes to be licensed as an operator of an SUV, 18-wheeler, or nuclear power plant—to develop skills that generally go unappreciated.

Lightplane pilots routinely multitask in ways that even a three-handed cell phone/Palm/GPS driver couldn't handle, and they can't pull over and park while they copy a new clearance from air-traffic control.

For instrument pilots, superb situational awareness becomes second nature, for if everything turns to excrement—which can happen—they're like blind men in a strange house who are suddenly told to go find the kitchen.

GA pilots operate totally on their own, without the emotional support generated by other small, cohesive groups such as infantry rifle squads, tennis doubles teams, racecar pit crews . . . or airline flight crews.

They are very good at compartmentalizing affairs, shutting out external concerns in order to deal with their immediate problems, whether it's a head-wind burning up fuel or unforecast weather blocking the only escape route.

The good ones don't tunnel-vision in times of crisis but actively assess and reassess their decisions—continually gathering information, refining solutions, working out alternatives. I realized that I would never be a superb pilot—an adequate one, yes, but never top-drawer—when I had an in-flight emergency and saw my concentration narrow to the size of a pencil beam, shutting out all rational thought and problem-solving processes.

I'd just taken off in a twin-engine Cessna 310 from Westchester County Airport, in suburban New York, for a short hop across Long Island Sound and the Island itself, to JFK to pick up a friend who had airlined in from Los Angeles. (This was back in the days when you could do things like that. Today it costs hundreds of dollars in landing and parking fees, plus a reservation made days in advance, if indeed it's possible at all.)

It was nighttime, but not so dark that I couldn't see the thick black stream of oil coursing out of the left engine's palm-size dipstick-access door as I climbed out of HPN. Oh, Jesus. My heart froze. The engine might seize in minutes. I'd still have a good right engine, but flying a twin on a single engine, at night, is a tricky and dangerous affair.

"Westchester Departure, I've got an emergency and need vectors direct Kennedy," I radioed frantically. For some inexplicable reason, I was at that instant desperate to fly over a large body of water and then another 20 miles to land at a busy airport at night with a failing engine.

"Uh, roger, Nine-Eight Quebec . . . would you consider a return to Westchester instead? It's only five miles behind you . . . "

Duh.

Flying an automatic Airbus on a regular route has its challenges, but it could be argued that they pale in comparison to single-piloting through bad weather to an unfamilar airport a turbocharged light twin. Aboard the latter, you're dealing with a cockpit forest of throttle, mixture, prop, turbo, and cowl-flap controls, dangerous single-engine handling qualities, and an analog autopilot, compared to the airliner's two power levers, two pilots, and near-totally automated, computerized operation. (Two things that come as a surprise the first time you fly a big airplane: they're way more stable on the approach to a landing than is a little one, and their size is really only apparent on the ground, when you're taxiing through close quarters. In the air, you're aware only of the cockpit, and a 747's is actually quite small.)

What is it about the person who can not only stick-and-rudder a Mooney 201 or a Cessna 310 with grease-it-on delicacy but can pound through lousy weather at altitudes that the airlines wouldn't touch, who can navigate with a tablecloth-size sectional, deal with icing using 1930s technology, and dart around thunderstorms with Kmart radar? And do it all without a first officer to put the gear down?

Good question, particularly since nobody has ever answered it.

Oh, those of us who fly little airplanes think we've answered it, and for decades we've traded personal insights into who's an airborne loser and who's the ace of the base. Consider our envious opinions of certain professionals. General-aviation pilots laugh that lawyers make lousy pilots because they don't realize that they're arguing not with a jury but with God. And doctors make lousy pilots because they think they *are* God.

Whatever parlor psychology we bring to play, these are obviously busy, preoccupied professionals who can afford both Mercedes-Benzes and the Benzes of airplanes. Which is one reason why the classic Beech Bonanza used to be known as "The Vee-Tail Doctor-Killer." MDs bought them when they should have been honing their talents on a safer, cheaper Skyhawk, the reasoning went, and soon found themselves in way over their heads.

In fact, the FAA has done an informal study that suggests there are two kinds of pilots: those who believe that fate is the hunter—that chance can kill them—and those who feel they are strongly in control of their destiny. The control freaks, and doctors and lawyers would most likely be in this group, seem to be at less risk of having an accident. Hmmm.

The general-aviation human-factors research that has been done deals largely with the man/machine interface—how best to communicate information to the pilot; what knobs, levers, analog gauges, and digital displays are most effective; how to help the human bean keep the machine right side up. Mapping out the physical, psychological, and demographic profile of the ideal private pilot wouldn't be of much practical use in any case. To wonder what makes an excellent Piper driver is as pointless as wondering what makes a great Ford Taurus driver. There is no mechanism for selecting or excluding either drivers or GA pilots based on skill sets, demographics, or personality profiles.

Nor should there be. In the U.S., anybody with a reasonable pulse and the means and the interest can learn to fly, and nobody much cares whether they're SuperPilot. We like it that way, even if we do occasionally try our best to fly over Long Island Sound at night with an engine dying. ✿

QUARTER-PINT

✿

BET YOU THOUGHT OFFENHAUSER ENGINES WERE OUT of production.

Not quite. There is one company on the planet that still manufactures the classic four-cylinder, twin-cam Offy full-race engine—an alcohol-sucking bantam that bellows through its headers with the same *whoopwhoop* challenge that rattled dirt tracks in the 1950s, when Roger Ward and Bill Vukovich rasseled little sit-up-straight Offenhauser-powered midgets sideways around the bullrings, cranking the nearly flat steering wheels like crazed bus drivers.

Maybe not the same voice, but a lusty little imitation of it. For these Offys are made by a small family-run Michigan Upper Peninsula machine shop called Replica Engines (www.replicaengines.com), and they are fully functioning, strong-running, compulsively accurate, quarter-scale versions of the original. Which makes them about the size of a beefy Nelson DeMille airport paperback stood on its spine.

"I grew up in a garage," says Wally Warner, who with son Scott, wife Betty, and daughter-in-law Evelyn runs Replica. "Learned to swear at an early age. It was during World War II, and at that point, cars were all prewar iron, and you often had to make parts to fix them, which is what my father did. I have a genetic disposition toward nuts and bolts, which Scott seems to have inherited," Wally laughs.

Piston rings and pins, valve springs and keepers, rod bolts, bearing shells, hose clamps, sump-plate bolts, crankcase ventilators, throttle springs, spark plugs and ignition wires, oil and water pumps, magneto, minuscule cam-cover fasteners, valve springs . . . every part the full-size engine had, so does the model Offy. Just way smaller.

Unlike typical model engines—the whiny little oil-in-the-gas, air-cooled two-strokes that power radio-controlled model airplanes and cars—the miniature Offenhauser is pressure-lubricated and water-cooled, which of course requires hugely complex, tiny oil galleries and coolant passages to be cast or drilled throughout the engine.

It's also not that much cheaper than a real Offy used to be. Fully optioned with a two-speed in/out gearbox, mounted on an engine stand that hides the 110-volt belt-drive pushbutton starter and the six-volt ignition battery, Replica Engine's baby Offy goes for $7,385 ($4,500 for just the engine).

Only a few Offys are left, for Replica makes its engines in limited editions of 100 apiece. They're also the world's sole manufacturer of fully functioning Ford V8-60 flathead quarter-midget engines; of 1937 Harley-Davidson Knucklehead vee twins; ancient Curtiss OX-5 V8s a-dance with exposed valve springs, rockers, and pushrods; and of World War I Gnome and Le Rhône rotary airplane engines, all of them quarter-scale. You could put each engine into a spacious raincoat pocket or clatter away the entire collection (which would cost you about $37,000 fully optioned) in a small suitcase.

"We started out selling the OX-5 for $4,500," says Wally Warner, "then we jacked it up to $5,500, finally to $7,500, and it probably should have been $10,000 the whole time. There's an awful lot of work in each of those things."

None of the Warners is getting rich. "We sell to people with deep pockets who have a considerable interest in things mechanical," says Wally. "There are lots of the latter but few of the former." And the combination is rare. "My brother is a John Deere dealer, and he insisted that farmers would be falling all over themselves to buy a working model of the two-cylinder engine that Deere used in its tractors in the 1920s and '30s, which makes a neat sound. So we did some research and yeah, there was great interest in buying them, but no interest in spending real money. We got three orders."

It's an unpredictable business. Harley-Davidson persuaded the Warners to make a run of 750 Knucklehead twins, since they were sure every Harley dealer in the country would want one. "Most dealers turned out to be too young to even know what a Knucklehead was," Wally admits, "and $3,800 buys a lot of profitable Harley T-shirts." So the Warners will be delighted to instead sell you a Knuck, which indeed runs with the inimitable loping exhaust note of a full-size Hog.

One Harley dealer who did buy the model put it on the parts counter, where so many customers fired it up with the little electric starter that it has been back to the Warner shop twice for valve jobs. (Plus a third service visit after some dork accidentally knocked it off the counter.) "If something like the Gnome rotary sees 25 or 30 running hours total, it's a lot," Wally says. "If you took real good care of it, you could probably get 40 before it needed an overhaul."

The unmarked shop, a small industrial building, is at the end of a dirt road that runs along the boundary of the sleepy little Schoolcraft County Airport. The U.P. of Michigan is one of the more depressed regions of the country—11 percent unemployment in an area that *The New York Times* dissed and dismissed as "the middle of nowhere" on the very day I visited the Warners. "It's really hard," grandmotherly wife Betty says. "Business comes and goes. But Wally and Scott seem to enjoy it, and we get by. We should be building something with no moving parts that we can sell for $2 apiece," she groans. "Like a hula hoop."

Wally and Scott, however, are moving in quite the opposite direction. Their newest engine, not yet complete, is dauntingly complex—a four-cam, 32-valve, dry-sump IRL Olds Aurora engine—and they're talking about doing a Ferrari V-12 and mulling the possibilities of a quarter-scale Rolls-Royce Merlin. The Aurora, with all the aural charm of a leaf-blower, seems an odd choice to model, but the Warners already have 28 pre-orders.

The Warners always need somebody to contribute a full-size donor engine to reverse-engineer, since they can't afford to buy them. They were fortunate enough to inherit an IRL Aurora that threw a rod during dyno testing, which trashed the block but left intact everything they needed to measure and miniaturize. And under a workbench, hinting at a future project, are the grimy pieces of a Chrysler Hemi that some drag racer used up, spit out, and donated. "Everybody loves a Hemi, even if they don't know what it means," Wally says.

Wally Warner started building miniature engines 26 years ago, after a tinkerer's career as a photo-lab owner who built all his own equipment, including an electrically driven 120-degree panoramic camera. Not surprisingly, he was also an avid r/c airplane modeler.

A company in Ireland, called Technopower, had been advertising a handsome radial model-airplane engine—not a scale version of any round engine in particular, just a unique little five-cylinder, four-stroke, air-cooled thumper intended for quarter-scale models of classic biplanes and the like.

Wally desperately wanted to buy one, but it turned out the company wasn't actually up to manufacturing and delivering the product. In fact, they were about to go bankrupt. So Warner bought the company, complete with its plans and tooling for the engine, moved it to the U.S., and soon was producing a line of Technopower radials with three, five, seven, and even nine cylinders.

The Technopower radials looked great, particularly in a scale Stearman or Beech Staggerwing, and they sounded even better. Most scale modelers were reduced to hiding a one- or two-cylinder off-the-shelf two-stroke behind a dummy radial-engine shell, which inevitably resulted in a Beechcraft, say, that looked like something Howard Hughes might have owned but sounded, when it flew, like a berserk Cuisinart.

Unfortunately, the Technos were substantially heavier and less powerful than the two-stroke hummers, and most modelers didn't realize that you had to fly an airplane equipped with one exactly as you would have the full-size version. You couldn't do a loop simply by snapping the r/c controller's joystick back; first you had to put the airplane's nose down, put some extra airspeed in the bank, and then work your way into a smooth loop with just enough knots at the top to keep from dishing out.

The Technopowers were never a success, but they did lead directly into the scale replicas: if we can make working generic radials for modelers who can't fly them, Warner figured, why not instead make working replica engines for the collector market that'll keep them on the shelf? (Some of his replica engines do end up in scale-model airplanes, cars, and even one model Harley motorcycle, but only a very few brave builders have actually flown them, sometimes with very expensive consequences.)

In the early Replica Engine years, the Warners did everything pretty much by hand, on lathes, drill presses, cutters, a screw machine, and an ancient reducing pantograph that they still frequently use. It is a classic Bavarian machine tool of awesome accuracy, with a finger that traces the shape of any complex three-dimensional object—an intake trumpet, distributor cap, rocker arm, oil cap, bell housing, cam cover, whatever—and simultaneously moves a whirling cutter that shaves the identical shape, exactly one-fourth the size, out of a block of plastic to be used in making a mold. Move the tracing arm and its heavily damped, precise, perfectly balanced path makes it plain that between input and output, any deviation will have thousandths in front of its number.

How accurate is it, I ask Wally? "Oh, it's dead-nuts. Probably half a century old. Been out of production for years."

Now the grunt work is done by a big Mitsubishi vertical machining center, Bridgeport Torq-Cut and Mazak lathe/turning center, all of them computer-controlled—a $145,000 nut that will take a lot of little engines to recoup. "The amount of the investment we've made really gives me the feeling we should be doing something else," Wally admits. "But Scott and I both enjoy doing this, though it's obvious the market for these replica engines will never be there to make us wealthy. These projects cost us a great many more hours than we ever imagined in the beginning. But we get started on these things and work all day, and hey, when you're havin' fun, you lose track of time."

As a small but highly competent machine shop, Replica Engines gets contracts for projects that range from the ridiculous to the occasionally profitable. They've done prototyping for the Argonne National Laboratory and the Naval Research Center (making devices so arcane or secret, they to this day have no idea what their purposes were), but they also attract the wackos.

As the 100th anniversary of powered flight approached in December of 2003, they were approached by one of the several serious builders of replica Wright Flyers, who sent the Warners superb blueprints of the original Wright engine and said he wanted them to build him one, full-size, to power his Flyer. No problem, Replica said, though Wally confides that "It was a terrible engine, don't know how it ever flew." They never heard from the builder again.

Replica also heard from an inventor who wanted them to build him a nine-cylinder radial motorcycle engine about the diameter of a pizza dish, but that never went anywhere either. And then there was the entrepreneur who intended to sell 10,000 keychains fashioned from miniature Gnome-rotary pistons. "The Mazak cranked out Gnome pistons for the longest time," Wally admits. "We delivered 5,000 and the guy ran out of money."

There is something indefinably magical about watching and hearing such tiny machines actually run. Perhaps it's a kind of minor megalomania, the ability to hold them in one hand, to control them, to own something in miniature that you'd never have in reality.

I've heard rotaries run at Old Rhinebeck, pilots blipping the ignition cutout to control their approach speed to the tiny grass strip—WHAAAAP . . . WHAAAAP . . . WHAAAAAP . . . the throttle on a Gnome stays either wide open or closed and can't be used to control airspeed—and Replica's model makes the same surprisingly strong, high-revs sound utterly unlike an ordinary ra-dial. I've sat behind an OX-5 in a Jenny with the late Frank Tallman, Holly-wood's premier stunt pilot, in the rear cockpit and watched the pushrods and rocker arms and valve springs clatter every which way, just as they do on Replica's version. ("Every time we run one we're scared something will hap-pen with all that stuff jumping around," Wally admits.)

And I'll bet that when the Aurora finally runs, it still sounds like a leaf-blower, just as it did when it won the Indy 500 four times in a row in the late 1990s. ✸

ALBATROS

✱

WARBIRD WANNABES STILL DREAM OF BECOMING SPITFIRE and
Mustang owners, but in the UK as well as the U.S., those with less money and
more sense are flocking to a far newer but somewhat less glamorous military
castoff, the Aero Vodochody L-39 Albatros. Not that there's anything truly
prosaic about a "Russian jet fighter," as the American media often refer to
the handsome, pointy-nosed L-39, but the fact is that the Albatros is neither
Russian nor a fighter, though incontrovertibly a jet.

The L-39 line is Czech, and the L-39C is a trainer, though it has hard-
points that will accept a variety of ground-attack ordnance and is operated in
that role by a number of third-world air forces as the L-39ZA. (There's also
an odd version called the L-39V, which is a C with a winch and over a mile
of gunnery target-towing cable in place of the backseat. Perfect for banner-
pullers with an audience of speed-readers.) The good news is that Czech
workmanship is vastly superior to the Iron Age efforts of the Russians—it's the
Czechs who make championship Zlin aerobatic airplanes, remember, not the
Ivans—and MiGs aren't that much fun to fly recreationally anyway. But the
best things about the L-39 are that it is available, it is economical, it is sup-
portable, and it is easy to fly. It's hard to say the same of Spits and Mustangs,
which those who own them know to be rare, maintenance-intensive and de-
manding of stickmanship.

"I know of two people in the U.S. who have sold P-51s and bought L-39s,"
says Quincy, Illinois, warbird dealer and L-39 specialist Dwight Barnell of Air

USA, Inc. (217–885–3800). "No, the Albatros isn't replacing Fouga Magisters over here, it's replacing the Mustang, if anything." The advent of the L-39 also put a dent in the American T-28 Trojan market. "When L-39s first became available, the bulk of sales were to existing T-28 drivers," Barnell avers. "T-28 prices range from $150,000 to $250,000, and 250K is the bottom side of the L-39 market."

Before its collapse, the Soviet Union operated about 2,000 L-39s as trainers. Today, those airplanes are virtually all parked, and the Russians can barely afford to brush the snow off them, much less maintain and fly them. Every 30 days, mechanics go out to flightlines all over Russia and turn over hundreds of L-39 engines . . . with 12mm sockets and six-foot extensions shoved up the engine bay from the tailpipe like shiny proctoscopes. The engines aren't even started.

The number of available L-39s varies as doors open and slam shut in this touchy trade area, but at times it has been said that the Russians were anxious to sell as many as 800 of their L-39s. So far, about 80 have come to the U.S., with 20-plus more in Western Europe and the UK. More are sure to follow, as the word goes out that the L-39 is cheaper to operate than are radial-engine and vastly slower, arguably uglier, and touchier North American T-28s. If you're brave enough to go to Russia and negotiate the deal for yourself, it's possible to get an L-39 for $150,000, or even less. Buy one from a reputable specialist in the U.S., and you can still get into your very own Mach .8 "jet fighter" for little more than $250,000—or up to $400,000 for a fully restored late model with new paint and state-of-the-art avionics.

"We've sold three L-39s in the last 90 days," said Barnell recently, "all for $300,000 with a single-navcom/transponder arrangement, no paint, no interior work of any kind. Those were mid-1980s airplanes. Some of the older L-39s—1974 to '76—they're running about $275,000."

Traveling to the former Soviet Union to do the deal yourself can be risky. If you have lots of time and the necessary talent to sit around a table and deal through interpreters with a bunch of people who hold all the cards, you can go to a Russian—or Lithuanian or Kurdistan—air force base, pick out the exact

L-39 you want, negotiate for extra spares or a zero-timed engine or whatever you wish, drink vodka with the middlemen from breakfast till late at night, make the necessary bank transfers, arrange for the shipping of the disassembled and crated airplane, and go home to await delivery, hoping that you might have saved as much as $100,000 on the deal.

Nothing will happen, however, until you've paid your money, and at that point, you're at the mercy of the Russians. Only when you crowbar open the shipping crate on your home ramp will you know if you got the 2,000-hour 1988 model you picked out, or a run-out '74 pig with bald tires and a corroded main spar. "So sue us," your new Soviet friends will say. And even if you *do* get the '88, you're liable to find—as a friend of mine recently did—that the Russians substituted an older APU for the prime-condition unit that had been in the airplane when he bought it, as well as a number of other, more minor substitutions. ("Sergeant, we need a new tip-tank drain valve for Number 36." "*Nyet* problem. Pull one out of the Amerikanski's airplane. He'll never know the difference. . . . ")

Such problems can be mitigated by a warbird professional, however. A good one will oversee the preparation and loading of the airplanes they've bought, or hire a company to do it for them. "If it's done correctly, you have an independent survey done through a company like Lloyds of London," Barnell explains. "They make sure the airplane has been correctly disassembled, that it is not damaged and is as represented, that it is secure inside the shipping container. Once the shipping company puts its tag on the container, it's sealed. But we also buy insurance, so that if the container shows up on my ramp and I get a box of rocks, I'm covered."

A pro will put funds in escrow or use an international letter of credit, and will ensure that nothing irrevocably changes hands until the right airplanes are officially stricken from the relevant air force's records and have safely passed through customs outbound.

"When the Wall first came down, we heard stories of MiG-21s being bought for $2,000," Barnell remembers. "Those stories were probably true, but that was just airplanes still on the air base. So now what do you do?

You've got to take the airplane apart, you need a bill of sale that'll be recognized by the FAA or CAA, papers that'll satisfy your government, you need to get it *off* the base, and arrange for transportation to the nearest seaport or across borders . . . there's lots to do."

Air USA not only takes care of all this, but also offers a variety of mods, including one that lowers minimum zero-altitude ejection-seat speed from 81 to 45 knots. They sell complete cockpit restoration and relettering in English, as well as turnkey L-39s ready to fly away. Also available: reflective gold canopy tinting for that F-16 look, or, if you wish, a "James Bond" pack of simulated Russian UB-16 missile launchers that can also be used as "a luggage compartment." (The airplanes that flew the fiery opening sequences of the film *Tomorrow Never Dies* were leased Luftwaffe L-39s. They too have gone west and are currently based in Minnesota.)

The key factor that makes the L-39 so desirable is that "It's the only available second-generation jet trainer," says Dwight Barnell. He explains that the warbird crowd has for years been flying warbird trainers, but that all are relatively crude turbojets with pedigrees stretching back as far as World War II. "The first-generation airplanes were the T-33, Fouga Magister, Pinto, Jet Provost, Soko Galeb, CASA Saeta HA-200, the Iskra, and the L-39's predecessor, the L-29 Delfin. All of them are 1950s and even '40s designs. After all, the HA-200 came straight off Willy Messerschmitt's drawing board."

The Albatros was designed in response to a Warsaw Bloc need for an airplane that had high, wing-shielded air intakes for freedom from foreign-object damage (FOD) when operating from unimproved runways; that put the backseat instructor up high enough to be able to see and correct what the "cadet" (as the Soviets called flight students) was doing in the front seat; that had large control surfaces and simple pushrods to move them, for small-aircraft feel that needed big-aircraft inputs; and that didn't need a lot of speed to do basic training.

So the L-39 was engineered to be a basic airplane that could operate off anything from concrete aerodromes to frozen lakebeds, with simple systems that didn't require high maintenance and weren't prone to mishandling by students. The equivalent American trainer, for example, would have complex hydraulic cockpit canopies that hissed up and down at the touch of a switch and locked into place automatically. The L-39's canopies flop open sideways and manually, like a Long-EZ's, and are locked closed with a hand-operated latch, just like a homebuilt's. And the entire windshield is hinged to fold manually up and forward, held in place by a simple prop rod, to allow access to the back of the forward instrument panel. Indeed, the core section of each panel slides about six inches aft on simple trackrods and then folds downward, for back-of-the-panel servicing, once a couple of knurled fasteners are released.

Another example: the ejection seats' primary parachutes—the ones that slow down the seat enough for the main chute to open without ripping—are each deployed into the airstream by a big but quite ordinary fishing-sinker lead weight that is flung up and aft by the charge of a shotgun shell. Simple, cheap, and reliable (as long as the pyro fires) rather than relying on some kind of high-tech mini-rocket or complex mechanism.

The ejection seats are rocket-powered. The straps are tightened and the chairs initially urged up their beefy rails by small pyrotechnic charges. They'll get you out at zero altitude as long as you have 81 knots of airspeed to aid in chute deployment. And yes, our FAA does allow warbirds to be operated with the seats hot.

The L-39 has thick, straight wings for lower landing speeds, vortex generators on the bottom of the horizontal tail to keep the elevator effective at those low speeds, and relatively large Fowler flaps. Though the big flaps would create large pitch changes as they deployed, they're linked to the trim system so that the airplane is automatically trimmed throughout the flaps' operating range.

Such cares were taken because L-39s were intended to train Soviet farmboys who'd never in their lives seen an airplane up close, much less sat in one. They weren't college boys or kids whose dads had owned Skyhawks, they weren't Embry-Riddle or CSE grads or ex-hot-rodders; they were fresh meat with little technological background. Yet they soloed L-39s in, typically, 12 hours *ab initio.*

"The Soviets wanted you to get all the jet-flight basics out of the way on this aircraft," Dwight Barnell explains, "because you went straight from it to a MiG or a Sukhoi, and *they* were real handfuls." (Many L-39 graduates also went on to helicopters, for the Soviets sent everybody, fixed- or rotary-wing, through the Albatros for basic training.) Therefore, much of the cockpit and panel layout and many of the instruments are identical to those in far faster Soviet jets. Operating the ejection seat, for example, is identical to the procedure used in a MiG-29, complete to the between-the-knees firing handles that the L-39 featured at a time when American front-line fighters still required an over-the-head reach—which can be impossible to do in a high-G situation. The U.S. Air Force and Navy have since gone to the Soviet system. Indeed, when our Lockheed F-22 next-generation fighter first flies, it will be equipped with a Russian Ka-38 ejection seat, one of a batch of 100 that the U.S. bought in much the same way that American warbirders are buying L-39s.

It takes about 10 hours of training, at a cost of $13,000 to $15,000, for a typical 1,500-hour complex-retractable private pilot to gain an FAA Letter of Authorization to fly the L-39. (The FAA demands a minimum of 1,000 hours total time before issuance of an LOA in any jet warbird, and there may soon be rulemaking requiring an instrument rating as well.) One outfit that does such training is Larry Salganek's Jetwarbird Training Center, in Santa Fe, New Mexico (505–471–4151, fax 505–471–6335, www.jetwarbird.com). Salganek has graduated dozens of pilots of widely varying backgrounds to LOAs in the L-39 Lockheed T-33, MiG-15 and MiG-17, Fouga Magister, and the L-39's predecessor, the L-29 Delfin.

The Beech T-34 Mentor, Salganek claims, is surprisingly similar to the L-39 in many ways. "The T-34 is a military airplane, it's got the same control feel, the same placement of the throttle and flaps, sensitive ailerons . . . it's a real similar machine," he says. "The L-39 is easier to fly than most airplanes in its category. It's got a real simple cockpit, it's easy to speed up or slow down—good throttle response—and it has all the qualities you look for in an airplane of that category."

Salganek likes his Albatros. "I've really been impressed with it," he says. "It's very dependable, very predictable, and very aesthetically pleasing. People like the way it looks, and they like the way it feels when they're flying it, and what else is it about?" he laughs. "Why else would you buy one of these things?"

He says, "It does everything a little bit better than anything else out there. It's a modern T-33—does everything the T-33 will do, just does it easier and on less fuel. It's going to be a great airplane as a warbird, definitely the best one of its kind."

The L-39 first gained notoriety in the U.S. in the hands of Chrysler Corporation co-Chairman Robert Lutz, a handsome, imperious, publicity-wise Detroit executive who, though he'd grown up in Switzerland, had done time in the U.S. Marines as a mud-mover—an A-4 Skyhawk pilot. When Lutz grew bored with driving his own company's sports cars and with flying (and once crashing) his MD-500 helicopter, he acquired an L-39, had it painted in authentic Soviet camouflage, and garnered lots of ink as the owner of what adoring motoring journalists generally assumed was "some kind of MiG." Like the Humvee quasi-military sport-utility vehicle in the hands of Arnold Schwarzenegger, the L-39 became a higher-profile warbird than it might otherwise have been.

Lutz got more publicity than he wanted, however, when he landed his L-39 gear-up at Willow Run Airport, in Ypsilanti, Michigan, his home base. Fortunately, he did it some 30 minutes *after* a Gulfstream-load of car-magazine writers had debarked and driven away from the Chrysler corporate jet in front of the company hangar where Lutz kept the Albatros.

It was the classic oh-shit scenario: Lutz had been distracted from his normal routine during a go-around occasioned by a dithering Bonanza on the active runway. He'd brought the gear up, turned a tight military-360 pattern to a close-in final, and never put the wheels back down. ("I'm gonna sue the tower. It was *their* fault," Lutz was heard to gripe days after the incident. Needless to say, he thought better of that plan.)

Lutz's biggest problem turned out to be the belly-slide's destruction of an unusual L-39 feature: a strontium 90–filled ice-accretion probe on the underside of the nosecone. Normally capped with a heavy lead shield on the ground, Lutz's was left in tiny, exceedingly radioactive pieces scattered along a Willow Run runway, and when Lutz admitted as much to the FAA incident investigators, the environmental-protection authorities showed up with a truck and crew that looked ready to clean up Chernobyl. Lutz got the bill.

What kind of performance can you expect from the standard two-seat trainer version of the Albatros—the L-39C? You'll rotate at about 95 knots, lift off at an immediately attained 110, and will need about 3,500 feet of concrete for safe accelerate/stop operation, and initial rate of climb will be about 4,000 fpm. Though the cockpit is lightly pressurized, you'll need to go on oxygen at about 23,000 feet; you can't put a high psi differential into an airplane that is designed to someday take a bullet through the canopy. Max level speed comes at 16,400 feet—405 knots—but a more typical cruise speed at that altitude is 340 knots burning 140 gph.

The L-39's prime operating altitudes are in the mid-20s, however, and service ceiling is just under 38,000 feet. "We like to use 22,000 to 27,000 feet, which gives us a hard deck at 10,000 for aerobatic maneuvers," Barnell says [assuming, of course, ATC clearance into a block of positive-control airspace]. "Our best aerodynamic and engine performance is at about 27,000. If we go any higher, we lose efficiency, since we are a straight-wing airplane."

Best-endurance altitude is about 25,000 feet, and experienced L-39 drivers plan on 1+ 45 cross-country legs with VFR reserves between refueling stops—about 400 to 500 nm, depending on the wind. Air USA sells a 70-gallon fuel bag that fits into an L-39's aft avionics bay once it has been emptied of its heavy, primitive, useless-in-the-U.S. Soviet radios, plus tip tanks modified to expand from 27 gallons to 61 and two underwing tanks that'll hold a total of 267 gallons rather than the stock 187. Fat with fuel, such an L-39 will fly for three hours, including VFR reserves. Or, if you want to gas to the max, there's the Air USA mod they developed for their U.S. Government adversary-aircraft contracts: a 110-gallon tank that fits in the rear cockpit, sliding neatly down the ejection-seat rails once the seat has been removed.

Landing, expect to enter downwind at 175 knots with a target speed of 120 to 140 knots on final, depending on flap setting. Bug speed will be 110 knots over the fence, then pull the power off and let the enormous trailing-link gear legs on all three wheels make you look good.

The gear is stout. One U.S. L-39 instructor tells the story of an arrogant new Albatros owner who came to him for "training." With the owner low and slow on one approach, the instructor took the airplane away, added bags of power, and made the runway. "This is *my* airplane," the owner raged after the landing. "Don't you *dare* do that again. Keep your hands to yourself."

Okay, have it your way. The instructor put his hands in his lap and sure enough, on the next approach the guy was again way behind the program. He landed well short, hit the six-inch lip of the concrete runway with an enormous bang, yet rolled out normally. "I climbed out, told him to go to hell, and walked back to the office," the instructor said. "But what amazed me was that there was absolutely no damage to the nosegear. On any other airplane, the impact would have wiped it out."

The L-39's Ivchenko turbofan engine puts out just under 3,800 pounds of static thrust and has a TBO of 1,000 hours—far below U.S. turbine standards—but then a newly overhauled engine can be had for PT6 dollars, typically $50,000 to $100,000 depending on the overhauler. A brand-new engine is $190,000, and there are plenty around. "When you're buying an airplane like this," Barnell cautions, "the number-one question to ask is, 'Where am I going to get parts?' Most first-generation jet trainers were operated by only one or two air forces. But the L-39 is operated by 36 different countries, which means 36 different sources of supplies. And it's still effectively in production, in the form of the L-59 and L-159 upgraded variants."

Which, by the way, are currently going new to various air forces for about $10 million apiece, to give you an idea of the intrinsic worth of a slightly used Albatros. No, the L-39 has no glorious combat history, and yes, it's slow enough to get rear-ended by a good Citation and doesn't even offer the baritone rumble of a Merlin V-12. But you'll have a hard time finding another legitimate "peacebird" that will provide anything like an Albatros's aerobatic capability (+8/-4Gs), sleek looks, safety, economy of operation, supportability, and reliability. The value of L-39s (and, unfortunately, the price of support parts) is going up daily. Besides, if you own one, you can legitimately walk the annual EAA convention's Oshkosh flightline in a G suit without looking silly. ✿

DB-601

THE ENGINE MAKES THE AIRPLANE. AN AUTOMOBILE, even a racecar, will run perfectly well with a variety of powerplants, configurations, horse-pressures, and displacements. But airplanes are different. They are simpler, purer vehicles.

Each airplane has a mission. The mission dictates the airplane's configuration. And the configuration requires a source of the right size, weight, and amount of absolutely, positively reliable power. Without a Pratt & Whitney JT9D, there would never have been a 747. Without a Rolls-Royce Olympus, no Concorde. Take away its Merlin and all you have is an Allison-powered A-36 Apache ground-attack light bomber rather than the ultimate all-around fighter of World War II, the P-51 Mustang.

In the 1930s, Americans were building airliners and prototype heavy bombers, and as a result became the masters of the round engine–the big air-cooled radials suited to the task. The Brits knew it wasn't Charlie Chaplin running loose in Germany, so they knew they'd need fighter engines far better than the lumpish 500-hp radials and in-lines that were then the norm. Contesting the Schneider Trophy for extreme racing seaplanes gave them excellent reason to begin developing huge 12-cylinder, liquid-cooled vee engines–ultimately the Merlin and, a few years later, the Griffon.

The Japanese were never contenders. Their famous Mitsubishi Zero was superlight and very maneuverable simply because it had to be designed around a little 780-hp radial engine that was Japan's best at the time.

The Germans, meanwhile, were barred by the Versailles Treaty from building fighters. So in 1934, Willy Messerschmitt designed the world's most advanced private plane, the Me-108—a 240-hp four-seat retractable that had capability, panache, and performance not equaled until the vee-tail Beech Bonanza appeared in Wichita after the war. Not surprisingly, the Me-108's basic design would prove suitable for a far more powerful engine. In 1935, the -108 was re-engineered to become the infamous Me-109, the fighter with which the Luftwaffe fought—and lost—the Battle of Britain.

(Rivet-counters are wailing—I can hear them—that the -109 should properly be called the Bf-109, since it was manufactured not by Messerchmitt himself but by a company called Bayerische Flugzeugwerke. Whatever. As long as we don't lose the best-known off-color joke featuring Queen Elizabeth. "How many Fokkers did you shoot down?" she asked an RAF ace during a medal ceremony. "Five," he replied, "all Messerschmitts."]

The Luftwaffe lost not because it had less-skilled pilots. The Germans had been training assiduously as glider and aerobatic jocks—activities that weren't forbidden by the Versailles Treaty—and some had even flown in hard combat during the Spanish Civil War. The RAF had to hurriedly snatch many of its novices from university flying clubs, a rather less formidable training ground. Nor did the Germans lose because the -109s couldn't handle the British Spitfires. The Messerschmitts fought at a great disadvantage, having to fly all the way from France, or even Germany, and arriving over hostile turf with, typically, five or 10 minutes' worth of gas to expend on dogfighting. The RAF pilots fought well within range of their own airfields and, if they were lucky, could parachute to safety even if their airplanes didn't make it.

Indeed, the Germans were within days of wiping out Britain's defenders when Hitler decided to concentrate entirely on bombing London—the Blitz—rather than destroying the RAF.

Until very recently, the engine that powered his Messerschmitts was virtually extinct from the earth, even while thousands of World War II Pratt & Whitney, Wright, Allison, and Rolls-Royce engines are still flying, driving everything from firebombers and midnight freight dogs to collector warbirds. Of the tens of thousands of DB-600-series engines that powered the Me-109, all manufactured in the decade between 1935 and 1945, only two remain operational, courtesy of the attention of a California engine restorer named Mike Nixon.

A peek inside a DB-601 gives a snapshot of German engineering sophistication. The -601 was fuel-injected, a German technology invented by Bosch for diesels, which we and the Brits knew little about other than as engines for submarines. Our motor-vehicle and airplane engines were all carbureted. When an Me-109 was too closely pursued by a Spitfire, the German pilot simply pushed over into a dive and thundered away while the diving Spit, also pulling negative Gs, unloaded its carburetor float, which flopped up and totally closed off the Merlin's needle valves. For several long seconds, the Spitfire became a glider. (Jaguar was still using carburetors when Mercedes put fuel injection on its 300SL racecars in the mid-1950s.) Admittedly, a DB-601's fuel-injection pump alone had about as many parts—nearly 1,600—as an entire British or American 12-cylinder engine, and had to be manufactured with the precision of a Leica. But never mind, that's the way German engineers like things.

Mike Nixon probably knows more about the DB-601 engine than anybody else who speaks English. Though his Tehachapi, California, company, Vintage V12s, specializes in Merlins and Griffons, he has now totally restored those two DBs as well, laboriously translating manuals, making tools, and using several donor engines for some parts and manufacturing others.

"During the Battle of Britain, that engine was a year or so ahead of anything the British or Americans had," Nixon says, "mainly because of the supercharger drive and the fuel injection." Allied superchargers were all gear-driven directly off the crankshaft. Daimler-Benz developed an elegant hydraulic drive for the -601's blower—essentially a miniature Dynaflow automatic transmission.

Here's another bit of ahead-of-its-time sophistication: an altitude-sensing aneroid controller varied the supercharger's speed so that all a Messerschmitt pilot had to do was firewall the throttle and dogfight. At least early in the war, American and British pilots had to carefully adjust the throttle as they climbed or descended to avoid overboosting the engine. A further complication was that most Allied superchargers were two- and even three-speed. You maxed out "low blower" at a midrange altitude, then manually shifted into "high blower" and began the delicate throttle-increasing process all over again.

"In the -109, you put the handle all the way up. It was a point-and-shoot airplane," says Nixon. "And the DB-601 is a very compact design. Their focus was on using a minimal amount of metal for the greatest amount of torque."

Nixon notes that the engine was full of inventive and interesting systems. The magneto timing was controlled by oil pressure, for example, and there was a lever in the cockpit literally labeled "spark plug cleaner." If a -109 pilot had been cruising at low power and found himself with oil-fouled plugs just as the P-51s showed up, he could retard the magnetos with that lever. "The combustion chambers would suddenly generate a whole lot of heat and burn the plugs clean; then, you could go back to full power," says Nixon.

The DB-601 was unusual in that it was an inverted engine. Its cylinder banks hung from the crankcase, with the crankshaft at the top of the engine. This configuration had a number of advantages. It allowed good visibility for the pilot of a single-seat, single-engine fighter, since the bulk of the engine was down low. The widest part of the engine—the two banks of cylinder heads—were down near the already wide midline of the fuselage. Mounting the heavier parts of the engine on the bottom also lowered the center of gravity slightly, leading to a better-handling airplane. And with the most complex parts of the engine roughly at a mechanic's eye level rather than stepladder-high, maintenance access was improved.

Which is odd, since Daimler-Benz wasn't happy about semi-skilled labor messing with its maxi-motors. Typically, the big V-12s were sent straight to centralized and specialized maintenance facilities for work if anything but the most minor service was needed. Wartime photos of USAAF mechanics often shows them, mere boys in baseball caps, standing atop 55-gallon drums and crude scaffolding, stripped to the waist with their arms deep in the entrails of an uncowled P-51's Merlin or pulling jugs from a B-17's Wright Cyclone as insouciantly as if they'd been back in Des Moines overhauling the Briggs & Stratton on Dad's lawnmower. The Luftwaffe's technicians were more likely to be wearing white shop coats and working at indoor benches.

Rumor has it that Rolls-Royce experimented with the same inverted configuration for the V12 that became the Merlin and that a visiting group of German engineers graciously was shown the prototype. The Germans went home and adopted the layout, and when Rolls learned that Daimler-Benz was going inverted, they switched to a conventional upright design rather than suffer accusations that they were mimicking the Germans. Rolls was always far too courtly for its own good in dealing with its ultimate enemies. In the late 1940s, as the result of an overly well-meaning trade agreement, Rolls shipped to the Soviets 20 Nene engines—just about the only jet engine in the world then worth a damn—and the Russians promptly copied them and completely re-engineered around the Neneski an airplane they were then designing. The result was the MiG-15.

The Luftwaffe used an early DB-601 to power the Me-209 to a world speed record and claimed that the -209 was simply an improved version of the -109. It was in fact no such thing, but was a special, one-off, lightweight record setter that could never have been turned into a fighter. Allied intelligence, however, fell for the fib.

An even better scam involved another DB-601 speed-record airplane, a Heinkel He-100 that was also far too highly wing-loaded and touchy to ever have become operational. Luftwaffe PR gave it the label "He-112U" to make it seem that it was simply a modified version of Heinkel's not particularly successful He-112 fighter. Then, after a 464-mph world speed record had been set, a flood of propaganda photos of the only existing "He-112Us" in the world, in a variety of different squadron markings and locations, was issued. The RAF was shocked, and it took several years for the truth to come out. It was the same few airplanes, all of them useless as war-wagers, and they had been painted, repainted, and painted again, and moved from one air base to another to pose as fighters, with Heinkel engineers playing the roles of pilots and mechanics preparing the planes for "missions."

Nixon's rebuilds are the only Daimler-Benz V12s flying anywhere in the world, both of them in totally restored Me-109Es. Which means that this exotic, short-lived powerplant, the ingenious machine behind one of the most fearsome weapons of World War II, is perhaps the rarest production piston engine you'll ever hear—*hear* being a critical word. If someday you're at an air show where one of Nixon's engines is cranked up and flown, it'll sound something like "a stonecrusher," as one ex-Messerschmitt pilot put it to English writer Michael Jerram. Jerram has himself heard a DB-601 in flight and describes it as "a rasping growl combined with the whistle of its hydraulic supercharger." (The airplane Jerram heard crashed in 1997, at the end of its very last flight before it was about to be consigned to an RAF museum.)

One of Nixon's restorations is in the hands of Microsoft co-founder and warbird collector Paul Allen (the other is in England), so the sky over Seattle is occasionally filled with the rumble of an engine that once helped to bring Britain almost to its knees. Beneath the wings of Allen's Messerschmitt, at such times, lie the vast factories of a company called Boeing, which once made thousands of B-17 Flying Fortresses, each powered by four magnificent Wright Cyclone nine-cylinder radials. Now there was an engine that made an airplane. And demonstrably did a better job of it. ✿

THE YO-YOS

✾

IF OUR MILITARY RAN WEAPONS-PROCUREMENT PROGRAMS the way
they did the Lockheed YO-3A project, the words "billions" and "dollars"
would never again have to be used by the DoD in the same sentence.

But then I'll bet you've never heard of the YO-3A.

The Yo-Yos and their immediate predecessors, called QT-2s, were the
original stealth planes, and they flew and fought for three covert years over
Vietnam and Cambodia. "We never had an airplane come back from a mission without having made hard contact with the enemy," says Les Horn,
then a young navy officer who played a huge role in the development of
what ultimately became an army project. "And I don't know of anybody
who took a round," adds George Walker, an army captain and YO-3A pilot,
who retired in 2000 as a brigadier general with 1,400 combat hours
logged, 400 of them in Yo-Yos.

The YO-3As flew at night and were essentially invisible, camouflaged in
a unique light/dark-gray pattern to hide them against a cloud-mottled, semi-
moonlit sky. More important, they were as close to silent as a powered airplane has ever been. "People told us they sounded like a flock of birds passing
overhead," says observer Mark Kizarik, a front-seater in the long-winged, bubble-canopied airplane.

"The pilot sat in back, because the observer needed the visibility," explains
Karl Forsstrom, another former observer. "In the army, pilots were all officers,
while the observers were enlisted men. Which created a lot of confusion when

we'd land at an air force base and some spec 5 would crawl out of what every-body assumed was the pilot's seat, with a captain in back."

I met Horn, Walker, Kizarik, Forsstrom, and a number of other Night Stalk-ers (as they called themselves) at the first-ever big reunion of quiet-stealth people, in a hangar on Meacham Field, in Fort Worth, Texas, last May. "Big," in the case of this deal, means a couple dozen guys and their wives. If you gathered together everybody who has ever flown or crewed a Yo-Yo, you'd need reser-vations for maybe 75. Including the FBI, NASA, and Border Patrol, and the Louisiana Fish & Wildlife guys who flew two surplus YO-3As after the war.

Yes, Fish & Wildlife. The Cajun cops were catching some 75 shrimp thieves a month by using their YO-3As' infrared sensors to spot the heat of outboards while the baffled poachers were wondering why so many birds were awake at night.

The YO-3As looked oddly like skinny-winged, high-performance sail-planes with propellers. Not surprising, since that's basically what they were—highly modified Schweizer 2-32 gliders with six-cylinder Continental O-360 engines in the nose, the entire airframe totally remanufactured, strengthened, flush-riveted, modified, and equipped by Lockheed.

Les Horn had gone to Vietnam as a young naval officer to see if it was possible to improve the effectiveness of the navy's riverine interdiction craft—including the kind of fast boats that John Kerry conned. "They were jet-drive, pretty fast, but you could hear them coming 15 klicks away," Horn says. The VC melted into the villages and forests long before Kerry and his guys ar-rived. "You want to know who Charlie is? He's the guy who just put down the baby and picked up the AK-47," Horn muses. "It was hard to find the enemy."

Horn first thought of using observation balloons. "They're very quiet, but they're also very hard to control," he quickly realized. He had a degree in physics and was also a navy pilot, so he started analyzing what produced noise on an airplane—the prop, engine exhaust, the mechanicals of the engine itself, surface cavities and protrusions, and form drag, mainly—and how to mask, attenuate, and dissipate it.

Oddly enough, a small group of engineers at the Lockheed Space and Mis-siles Division had been thinking along the same lines, and in one of the ideal matchings of talent and task that the military occasionally achieves, Horn was plucked out of the paddies and plugged into a DARPA (Defense Advanced Research Projects Agency) assignment at Lockheed.

Horn knew that the area to concentrate on was muffling mid-frequency noise, where the human ear is most sensitive. Low-end noise travels a long way with relatively little attenuation but is hard to hear until it is a lot louder (more sound pressure, or dBs) than mid-range noise, and we all know that only dogs can hear very-high-frequency noise. Horn also found that just a small amount of background noise—insects, a little wind in the jungle trees at night—could mask the residual noise of a super-quiet airplane. (How? Well, like somebody turning on a bathroom faucet so the dinner guests won't hear him peeing.)

Lockheed built on the cheap an ungainly proof-of-concept vehicle called the QT-2, a Schweizer 2-32 sailplane with a little 100-hp flat-four Continental engine in a soundproofed box atop the fuselage behind the pilot. QT supposedly stood for "Quiet Thruster," but the designation had to come from somebody who knew what the phrase "on the QT" meant.

A propshaft driven by rubber belts that both eliminated gear- or chain-drive noise and substantially reduced crankshaft-to-propeller rpms ran awkwardly above the cockpit. ("Only full-size aircraft in the world driven by rubber bands," Horn points out.) The shaft stretched to a four-blade wooden prop atop a big pylon on the very nose of the airplane. A pylon that was, in effect, an extra vertical fin, which did little to enhance the handling qualities of the already-awkward QTs.

The props—several were tried, including a six-blader—were handmade by a man named Ole Fahlin, an ex-World War I Swedish Air Force veteran whose reputation had been established not in the military-industrial complex but as a supplier of custom wood props to homebuilders and formula air racers. Fahlin would come out on the Lockheed ramp between test flights of the original QT-2 and tune the prop by eye with a common wood rasp—take a little off here, reshape the blade there . . . "I don't know how he did it," Horn admits.

Most of the people involved in the QT program were old-line Lockheed craftsmen, fabricators who did a lot of the last-minute work at home, often overnight. One had a pickup truck with a cap, and one night, security stopped him at the gate. What have you got back there, they asked. "A whole lot of stolen aluminum," he replied.

"Aw, get outa here," the guard laughed.

So he drove out of the gate with his load of aluminum stolen for the QT.

Two of the QTs were sent to Vietnam for a highly successful operational test, but the 210-hp, retractable-gear YO-3As that followed them were the surveillance platforms that the army really wanted—fully operational, 3,100-pound-gross airplanes strong enough to carry a reasonable fuel load, lots of electronics, and a crew of two under a big bubble canopy. (Schweizer 2-32 sailplanes weighed just 1,400 pounds at gross.)

Development of the YO-3s was so secret that they were tested at night, at a remote Nevada airstrip that had once been a World War II B-24 training field. Not remote enough, however; the Lockheed crew had to take down the red airstrip beacon because whorehouse customers kept showing up on their way to the nearby Mustang Ranch.

The Yo-Yos sought their targets with infrared spotlights and a light-amplification system similar to today's Starlight system, and the observer would peer through a big hand-held scope, about the size and weight of a roll of steel paper towels, and try to keep from barfing. While the airplane banked and orbited in the dark, he had one eye on the ghastly, artificial bright green of the optics and the other on the instrument-lighting redness of the cockpit. He was looking for campfires, cigarettes, truck exhausts—anything hot.

"We found a lot of sampans," George Walker says. "A soldier is a soldier wherever he is, and he'll light a cigarette or start a fire if there isn't somebody right there to tell him not to." The VC seemed particularly vulnerable to the chill of the night and fired up little charcoal heaters on their sampans as soon as the sun went down.

"We had two tasks," Karl Forsstrom explains. "One was simple reconnaissance—go out and see stuff and bring back information. Or if we had enough fuel aboard and the target was important, we'd hang around and call in artillery." Yo-Yos also worked with an Army Huey gunship squadron that had infrared capability. The YO-3As would find the target and call in the gunships. The VC would hear the Hueys from miles away, of course, and put out their fires and other heat sources, but by then it was too late. The YO observers would light the bad guys up with their infrared spotlights, "and the Huey gunships would come in and engage the target on their first run," Walker says. "We were very good at that."

Few are aware, since the silent-surveillance program was a secret for decades, that the YOs were the first aircraft in the world to carry laser target designators—the systems that today guide smart bombs to their targets. "They didn't work very well, maybe didn't work at all, but it was a beginning," says Sherman Seltzer, Lockheed's chief engineer on the quietplane project. "It proved the ideas that we've refined since then, we just didn't have time to refine them on the YO-3A."

The Yo-Yo's laser beam was sent down through the observer's Starlight-scope optics, a process that weakened the beam enough to make it ineffective above a relatively low altitude. Nor did it help that there weren't yet any weapons in the inventory to use laser guidance.

Still, George Walker remembers one time he actually employed the laser system. "They wanted us to go out and find a good target at night, mark it with the laser, and call in a Cobra gunship," Walker recalls. "We finally convinced them to do a daytime test first, on an artificial target. So I lit up a wrecked APC [armored personnel carrier] and here comes the Cobra. Since they were going to roll in hot, I wanted to make sure we all understood exactly what was going on, so I called them and asked if they had acquired the target and were sure exactly where and what it was.

"'Affirmative,' they answered me. 'We have the target in sight. A two-tone, gray-camo, single-engine airplane.' It sounds funny now, but it didn't seem amusing at the time," Walker recalls.

Dick Osborne flew what was probably a YO-3A's most controversial mission, when a North Vietnamese trawler had been tracked from Haiphong to waters off southernmost South Vietnam. It was obviously there to unload supplies for the VC, but the navy wanted to know exactly when and where it would do so.

Osborne was ordered out on a night surveillance mission—illegally, it would turn out, since it was an area of operations strictly limited to the navy—and says, "We saw at least a hundred small and large sampans congregating in an area where we usually saw none." As he watched, the sampans fled while the navy surrounded the trawler and blasted it both from the surface and the air.

The trawler was the sole North Vietnamese steel ship sunk during the entire war, though in fact the likelihood is that the crew opened its seacocks and intentionally scuttled it when a Navy OV-10 Mohawk hit and disabled its rudder. Whatever the case, it was the biggest single target ever acquired by a YO-3A.

Ultimately, the Yo-Yo program became a forgotten backwater. Of the 14 airplanes sent to Vietnam, by mid-1972 only five were left. The others had all been lost in accidents of one sort or another, though none to enemy fire. "The war was winding down," Walker recalls, "and we'd found that 85 percent of our sightings had been initiated by the pilot, who had then directed the observer and his nightscope to it. So we could have achieved pretty much the same results by taking a well-muffled single-seat airplane, given the pilot a Starlight scope, and done away with all those sophisticated electronics. We'd proved the concept—that you can find the enemy through a system of stealth— but it simply wasn't that good a system at the time that we could convince the public we should spend more money on it."

In 2004, 32 years later, the DoD published a Request for Proposal that went to Lockheed, Schweizer Aircraft, and other potential contractors. It asked for the development of a manned, propeller-driven, low-altitude, low-cost, quiet, covert reconnaissance aircraft to use in situations where high-tech radar stealthiness is not required. Against Iraqi insurgents, say.

Unfortunately, it was really too late to go back to Schweizer and ask that the 2-32 sailplane production line be reopened so 21st century Yo-Yos could be built. A lawsuit put the company out of the airframe business when the wife of a 2-32 backseat passenger sued after he was killed in an otherwise minor landing accident because he'd unfastened his seat belt to shoot video over the shoulder of the front-seat pilot, who was unaware that he had an unbelted rider behind him. ✹

The Grandest Piano

✻

The Steinway Model D concert grand—a combination of musical instrument and fine furniture that weighs almost half a ton and stretches a quarter-inch shy of nine feet from keyboard to prow—is one of the most complicated handmade machines ever built. The only thing that might trump it is the modern Formula 1 racecar, but despite its considerable degree of hand manufacture, a fair number of a Mercedes McLaren's or Williams BMW's parts still must be machine-made.

Many people assume that Steinways, those paragons of Old World craftsmanship, are made in Dusseldorf, or perhaps Paris. Yet Steinway is, and always has been, a New York City company. For over 150 years, they've been turning out pianos first in Manhattan and now in Queens, right down the street from the "Ferrari Driving School." (Though its logo is a bright red on fly yellow replica of Enzo's, it's in fact a truck-driving school miles from the nearest 360 Modena.)

Henry Steinway invented the piano as we know it. Oh, sure, there were pianos for centuries before he came along, but they were made entirely of wood, and their sound was tinkly and sparse. No wooden case could possibly hold strings wound tightly enough to resonate powerfully. Inside a Steinway D, the 243 tempered, hard-steel strings stretched across the piano's harp—the huge hand-cast iron framework at the heart of the instrument—exert 35 tons of pressure. Which is probably enough to crater a three-bedroom house if strung between attic and cellar and tuned tightly enough to play "Chopsticks."

Steinway, an engineer, had a close relationship with Hermann von Helmholtz, one of the widest-ranging of 19th-century scientists and generally considered to be the father of the science of acoustics. Between von Helmholtz's informal input and Steinway's understanding of stress engineering, the modern piano was born more of science than art. It was, in fact, a direct creation of the Industrial Revolution.

For decades, the Steinway factory was operated much as it had been in the late 1800s, and it was said even in the early 1990s that if a Steinway worker from those days had been resurrected and shown how to use a Metrocard to take the subway to Queens, he or she could have punched in and gone straight back to work.

There are still remnants, throughout the factory, of the era when all of the thumping, clattering machinery was belt-driven from a big master shaft. One of the oldest is what they literally call "the pounding machine." An arrangement of cams and plungers designed and built well over a century ago, it breaks in each new Steinway by banging every key on the piano 8,000 times—over 700,000 notes struck—in 45 minutes. No, it doesn't make music, it makes a random cacophony fortunately only dimly heard through the soundproofed walls of its workroom.

But Steinway has more recently become a tech-savvy company, though with a difference. "Technology is an area greeted with great skepticism around here," says Steinway's Director of Quality, Robert Berger. "We give very careful consideration not to the increase in productivity but to the quality level that it produces when we put in a piece of modern equipment." Steinway doesn't need more productivity. The company makes fewer than 5,000 pianos a year, and they don't have the space or skilled workers to make any more.

So in one corner of the factory, you'll see five or six brutes muscling a thick, 22-foot-long plank of rock-hard, 18-layer laminated maple into an enormous 100-year-old clamping jig invented by a Steinway son. They are using huge pry bars, pneumatic wrenches, heavy mallets, and thick steel clamping bands to make a Model D's sinuous and acoustically crucial one-piece rim.

It's a calm and practiced but no-time-wasted ballet, since the glue, which initially acts as a lubricant to let the layers slide against each other, begins to set up solid in 20 minutes.

Yet not far away, a numerically controlled Shoda NCW-516 numerically controlled router is quietly cranking out hundreds of action parts the size of sugar cubes, each one requiring a dozen separate automated cuts.

The level of accuracy that Steinway attains in such operations—plus or minus .001 inch—is no big deal for the metal-milling business, but for woodworking, that's phenomenal. It does require perfect wood, however: Steinway's acceptance standards for the Sitka spruce that it buys are one level higher than aircraft grade, where rare Alaskan spruce is still used for some crucial structural parts, particularly in kitbuilt airplanes. (Steinway once made airplanes, in fact. During World War II, they built Waco troop-carrying gliders, some of which landed in Normandy on D-Day.)

Some of what Steinway does, however, still relies purely on human hands and eyes rather than laser aligners or computer-controlled milling. The worker who performs the final adjustments of the action does it entirely by feel, shimming the mechanism to perfection with tiny pieces of paper.

My brother Leland, a designer of advanced statistical software, is one of a tiny group of nonprofessional pianists who own Steinway Model Ds. For an amateur to own a D to play for the fun of it is like commuting in a John Force funny car. "You have your Porsche," Lee laughs, "and I have my Steinway." He's right. At $85,000, his D cost him exactly what a well-optioned 911 Carrera goes for. (The most expensive new Steinway ever sold was a lavishly decorated, hand-painted "art piano" concert grand that in 2002 went out the door for $675,000, but the all-time record is held by John Lennon's battered little Steinway upright, auctioned for $2.1 million.)

Ancient Strads and mellow Martins are considered desirable, but that isn't the case with pianos. "If you're a serious player, you're going to want a brand-new one," Leland says. "If you play it every day, which is the case when you're dedicated enough to have a Model D, a piano will wear out in 25 years. You can throw away the action, the strings, the soundboard, and pretty much start over again to rebuild it, which Steinway will do for you, but no professional would say, 'Oh, I'll take the old one.'"

Like a high-performance car, a Model D needs careful servicing. "You need to have it tuned probably four times the first year and twice a year after that," Leland says. "And piano tuning is not a matter of listening to a tuning fork and deciding whether a string is in pitch; it's extremely technical. You have to recognize the difference between seven, eight, and nine beats per second and figure out which intervals have which beats. There are books on tuning, but it's not something you can teach yourself."

My brother is a hands-on guy who oversaw the construction of his equally rare Dinan-ized and M-engined BMW 735 track car, and he understands every nuance of the workings of his Steinway's action—the enormously complex, Steinway-invented system of arms, levers, pivots, hammers, jacks, and keys, exactly 5,104 wood, felt, leather, steel, bronze-alloy, and phenolic parts per piano, whether it's a studio upright or a concert grand. (Steinway makes Models S, M, L, B, and D. Whatever the letters stand for, the standard joke is that they mean Small, Medium, Large, Big, and Damn Big.)

"What amuses me," Lee says, "is that professional pianists generally have no idea how a piano actually works. They simply use them as tools. They're like race-car drivers who understand exactly what they need to know about camber and spring-loading and tire pressures. They can tell a crew chief exactly what's wrong with a racecar, but they don't have any idea how or where to do the adjustments themselves. Professional pianists are like that. They don't care what a piano looks like or how much it's worth; they just want to feel and hear."

Is a Steinway D the best piano in the world? Other manufacturers have tried to snatch that reputation away by making more expensive pianos. Bosendorfer was briefly touted as "the best" largely because their concert grand was more than twice the price of a Model D. "I once played a Bosendorfer Grand Imperial, their legendary piano," brother Leland recalls. "It was sitting next to the Steinway I bought, in the showroom, and I tried them both. I was really surprised; it didn't have the Steinway's rich tone. It had 12 more keys in the bass, but they're never used. There's no written music for them other than stuff by a few minor composers who were commissioned to do something you could only play on a Bosendorfer."

Yamaha even developed a hand-built, enormously expensive concert grand that had nothing in common with the company's production pianos, but they could only get a single concert artist (Andre Watts) to use it. Considering that fully 99 percent of all active, major, professional piano drivers prefer a Steinway's horsepower even though there are 15 other manufacturers who would gladly give them pianos for free, this seems to be one race the Steinway has won. ✿

THE JOY OF FLYING

✵

WHEN I WAS A KID, FLYING WAS the most exciting thing you could do. *Extreme sports* meant riding your Schwinn down a steep hill and hoping the coaster brake worked. Adventure travel was going to Cape Cod with your parents. And a hot car was a 58-horsepower MG TD.

But the rare airliner trip . . . my very first flight took me from LaGuardia to Nantucket Island, a 45-minute ride in a Convair 240 that I imagined was a Martin B-26 Marauder with windows. I still remember the buckety cough of the R2800 radials, the very engines that only a few years earlier had powered P-47 Thunderbolts and F4U Corsairs at war. Idling, they loped and chuttered like a fleet of old farm tractors.

At the end of the runway, the captain held the brakes and ran up the big Pratt & Whitneys, one after the other, a torrent of sound, every part of the airplane shaking and rattling as the tires skittered, unable to entirely restrain the power. My worldly dad told me they were "checking the mags," whatever that meant.

MmmmWHAA . . . *mmmmWHAA* went the props as the gold-striped supermen up front cycled them, in what seemed a strangely fatalistic hope that if at least the basic systems worked, perhaps we'd survive the trip. Then, like a sprinter bursting off the blocks, the Convair bolted down the runway—at least, so it seemed to me—in a whirlwind of prop blast, 36 pistons the size of a kid's beach-sand buckets flailing away inside the oil-streaked yet shiny nacelles.

So different than the electric-train hum and steady takeoff torque of today's efficiently packaged winged mailing tubes. Ten-wide ranks of passengers sit placidly, staring at newspapers or nothing, in a flying cineplex waiting for the movie to begin. Back in the 1950s, slipping the surly bonds was a noisy and tentative process. Today, we don't even bother to look out the window.

How far we've come . . . or have we? My Convair ride took place exactly half a century after the Brothers Wright launched their shaky bird and Orville somehow managed to survive that first powered flight in an airplane almost as unstable as an F-16. He was the first of many early birds whose introductory flight was simultaneously their first powered solo. Serious on-the-job training.

Only 50 years after he managed that, I rode a 4,000-horsepower airliner that had the cocky legend RADAR-EQUIPPED by the cabin door. Wow! What will they think of next? Boys dreamed of becoming airline pilots—it was too early for girls—and I can still recall the Ford ad that showed nothing more than a car's overhead cabin-light switches and a hand in a four-striped sleeve reaching up to activate them. The gold bands said everything about what kind of man drove a Thunderbird.

One of my first assignments as a young magazine staffer was to write about a senior Pan Am 707 captain who commanded various legs of the line's famous around-the-world flight, PAA 001. He ate then-exotic sushi in Tokyo, overnighted in Rangoon, took time out to visit the Pyramids, partied in Rome, stopped in London to be fitted for bespoke shoes on Bond Street, and after finally landing at what was then called Idlewild, in New York, he commuted home. To Bermuda. What a life.

Being a baby editor, I wasn't permitted to actually accompany the skygod. A big-name photographer did that, and I wrote from the f-stop's notes. But I did get a 707 ride to Bermuda and immediately back, courtesy of Pan Am, so the kid could get a taste of what jet travel was like. It was 1961's version of the Concorde, and I spent most of the time looking out the window in awe at the Atlantic, seven miles below.

There is a way to close that seven-mile gap, to bridge the vast difference between aviation as mass transit and the real thrill of flying. All it takes is a windshield instead of a porthole—a front seat instead of a window seat. Only then can you see, sense, and enjoy flying in all its dimensions.

Peter Garrison taxis his one-of-a-kind single-engine airplane *Melmoth 2* into position on Runway 12 at Whiteman Airport, in suburban Los Angeles. *Melmoth*'s wings are as long and narrow as bread knives. They seem far too insubstantial to hold aloft the large, bubble-canopied fuselage. Like no other four-seater currently flying, the airplane is configured with two conventional front seats but a pair of rearward-facing seats in back, an ergonomically artful piece of passenger-packaging under a Plexiglas solarium. It allows the canopy to have a proper airfoil shape, since the occupants' heads are all clustered in the middle.

Garrison holds on the centerline briefly as he carefully pushes up the power. The 210-horsepower Continental's turbocharger surges and settles down, and he releases the brakes. Garrison flicks a glance at the oil pressure, rotates at 75 knots, and brings up the gear. I am in the right seat, next to him, and a fat aluminum torque tube turns slowly next to my right ankle as it activates the nosewheel. On it, the Penteled legend GEAR UP rotates into place on the tube.

We're climbing out at . . . well, it's hard to tell. The vertical-speed indicator is pegged at its maximum of 2,000 feet per minute up, and *Melmoth* is leaving astern its twin-engine chase plane as though that Beech Duchess were a rusty Camaro blown away by a Dodge Viper.

Our fiberglass seat shells are unpadded. Peter sits on a tiny throw-pillow; I perch on a sheet of something a step up from bubble-wrap. The cockpit is bare, the airplane's reinforced-plastic structure naked. Through a small area where someday a flap-actuating mechanism will pass, I can see the ground rushing by below, like a road seen through the lacy floorboards of an old VW.

Holes in the floor? Penteled legends? Unpadded seats? What's going on here?

What's going on is that *Melmoth 2* is placarded "experimental," a homebuilt. It's also unfinished, a work in progress, and Peter and I are taking off on what is in many ways a test flight. "This is the most boring part of building an airplane," Garrison admits. "Testing and refining all the systems so they operate properly for more than 20 hours at a time. The fun part is when you're working with lines on paper, and every line is filled with so many possibilities."

Boring for Garrison, perhaps, after 20 years of designing and building, redesigning and rebuilding. But for me to be aloft, even simply as a passenger, in the creation of a man who for 35 years I'd known as a friend and deeply admired as a writer (Garrison's day job) is thrilling. Peter and I go back a long way.

Think back to what you were doing over 20 years ago, in 1983. Fooling around in high school? Grade school? Just married? Climbing the corporate ladder? That's when Garrison began to build *Melmoth 2*, in a one-car garage along a steep LA backstreet. An old shack so narrow that passengers would have had to debark before a car could be parked inside it.

With the help of his friend Burt Rutan, the plastic-airplane pioneer, Garrison learned the craft of laying up and vacuum-bagging carbon-fiber-reinforced composite—a process in which the resin-wet plies are laid up draped not only into the mold that will shape them into wing panels or fuselage pieces, but at the same time, inside a big baggie. When the air is sucked from the bag with a vacuum pump, it squeezes the layup tight and forces out all the excess resin.

Simple, heavy layups of fiberglass cloth slathered with epoxy and left to settle are fine for boats and Corvette parts but anathema to an airplane. In aviation, one classic but rarely satisfied test of what to add to an airplane under construction is to hold the part at arm's length and let go. If it rises, it's fit to fly. If it falls to the floor, leave it off.

Garrison admits that making *Melmoth* out of carbon-reinforced composite rather than the classic sheets of riveted aluminum saved little if anything in weight, and cost considerable construction time. "And there's always the fear that you're not doing it right," he says. "With layups, your mistakes are hidden inside the composite, while with metal, you can see if you're doing it wrong."

Composite construction did, however, allow him to "loft" the airplane's shape as whatever combination of compound curves he and the wind wished, rather than simply defining traditional straight-line bulkheads, stringers, and longerons familiar to any model-airplane builder. To do that, he created a complex computer program called Loftsman.

Take a look at Loftsman at www.aerologic.com, where you'll also see a computational fluid dynamics (CFD) program called Personal Simulation Works that Garrison and a partner wrote. CFD analyzes the effect on a "fluid"—in this case, air—of a body moving through it, numerically indicating where the fluid creates various degrees of lift or drag, friction or pressure. The beauty of Garrison's PSW is that it reads out in vivid hues rather than stark numbers, flowing over a digital wire-frame representation of whatever shape is being tested. The wind in living color.

Wilbur and Orville would have been stunned to see Loftsman or PSW, but they still were the world's first "homebuilders," as the aviation world calls craftspeople such as Peter Garrison. The Wrights, serious hobbyists at heart, designed and entirely on their own built a handmade, one-of-a-kind powered airplane *and* its engine. We call it the Wright Flyer.

They were the first. A century later, 60-year-old Garrison has become the most recent of that rare breed—serious-hobbyist designer-builders.

There are three kinds of homebuilders. Most of them (including me, when I built a two-seat Sequoia Falco) assemble a kit—an existing, proven design that UPS drops on the porch as boxes of parts plus a fat some-assembly-required instruction manual.

Other more ambitious homebuilders start with blueprints, again of tested designs, and do all of the fabrication as well as assembly on their own—an enormous job.

Yet a precious few not only fabricate and assemble but also design and engineer their own flying machines. Garrison is part of this tiny club. Wilbur and Orville started it.

Peter's airplane is called *Melmoth 2* because there was a substantially different two-seat, Garrison-designed *Melmoth 1* that flew throughout the 1970s. Flew well enough, in fact, to cross not only the Atlantic but the entire Pacific, as well as nearly the length of South America, feats far beyond the capability of factory-built Cessnas, Pipers, and Beeches, unless they had been outfitted with special "ferry tanks" and, typically, flown solo.

That original *Melmoth* was destroyed in a freak accident, when an out-of-control Cessna ran into it at an airport and Cuisinarted the airplane with its spinning prop. "Everything that I built was destroyed," Garrison says. "Everything that I bought off somebody's shelf—the engine, the avionics, the instruments—survived."

Survived to become part of *Melmoth 2*, a remarkably advanced flying machine that is the product of one man's invention, intuition, imagination, and engineering talent. Engineering talent? Garrison's only degree is in English, and not even from MIT but granted by its Cambridge neighbor, Harvard.

As an aerodynamicist, stress analyst, structural engineer, software writer, loftsman, draftsman, computer-aided designer, fabricator, and test pilot, Peter Garrison is entirely self-taught. So were the Wrights.

Over the San Gabriel Mountains, north of Los Angeles, Garrison cranks *Melmoth 2* through a series of steep turns, the barren hills sliding past not far below. I remember how my shoulders would ache after a flight, when I took flying lessons in the mid-1960s, because I was unconsciously resisting the feeling of sliding out of the airplane, of falling, that sometimes overcomes novices when they are inside nothing more than a lightplane's airy shell of thin Plexiglas and aluminum skin.

The sensation soon passed, replaced by a pilot's pleasure at being able to make the world tilt and revolve, to displace the horizon and reverse the natural order of up and down in a way not possible in any other human activity. There are pilots who apparently enjoy the gut-rattling, face-warping, brain-mushing, slam-and-bam moves of high-G competition aerobatics, but I'm not one of them. Better the balletic swooping of a lazy eight, a loop that leaves you briefly weightless at its zenith, a slow, spacey barrel roll. Or hey, I'll take the steep turns and let it go at that.

Garrison is a realist, not an enthusiast. (Again, so were the Wrights.) There are a hundred things he'd rather do than fly, and he creates airplanes not to indulge in the new private pilot's routine of boring holes in the sky, as they self-deprecatingly call it, or of flying pointlessly from airport to airport to eat $300 hamburgers, which is the cost of such a meal when avgas and other operating expenses are factored in.

Peter began designing and building airplanes because he wanted to go places. Around the world, originally—one flight he never got a chance to make in *Melmoth*, but perhaps will in *Melmoth 2*.

The original *Melmoth* had two seats, for Peter and his companion Nancy Salter. *Melmoth 2* has space as well for their son and daughter, 22-year-old Nicholas and Lily, 15. Yet none of them has the faintest interest in flying, and Garrison might just as well have built a single-seater. The joy of flight is not for everybody.

"But Nick was deeply affected when the airplane flew for the first time," Garrison says, "I thought maybe it was because it showed that his father wasn't a deluded crank after all." In fact, it was probably because *Melmoth 2* had been in the background—and sometimes the forefront—of young Nicholas's life literally forever.

If Garrison is brave to fly untested, amateur-designed airplanes, Salter, a fan of the airlines, is indescribably fearless. Crossing the Atlantic behind a single engine, she read a book the entire way. Over the Pacific for 15 hours, she mostly slept. Above the impenetrable jungles of South America, Salter at one point seriously considered having to use the parachute she wore, when Peter got lost and ran *Melmoth* perilously low on fuel.

Shadows stretch east at the end of the day as we hurry back to Whiteman in a long, smooth, power-on descent among scattered clouds. *Melmoth 2* doesn't yet have nighttime nav lights, for they're far down Garrison's to-do list, which currently includes some 51 items, from "install longitudinal restraint for turbo" to "fix shimmy-damper leak."

We beat the sunset, the mainwheel tires squeaking briefly as Garrison holds *Melmoth 2* off the runway for a gentle, nose-high touchdown. As we taxi back to his hangar, pilots standing in small groups outside their own hangars turn to watch, all of them knowing they are seeing something very special—an airplane that is the product of a single individual's perseverance.

Ultimately, the joy of flying will exist forever in those people who can't resist turning to watch an airplane taxi past or fly overhead. I still do it and will never stop. For there is something still magical, still inexplicable about flight. Rocketry is no more complicated than a flung stone or a shot bullet, but how an airplane's wings cancel gravity remains a mystery even though I once thought I understood Bernoulli's Principle. (Peter Garrison once admitted to me that he had his own epiphany one day when he got stoned and was able to envision air molecules as "tiny baseballs," but it hasn't worked for me.)

I'm sure there's an easy answer, but I don't want it. I prefer flight to remain fabulous, fascinating, mysterious . . . inexplicable yet endlessly joyful. ✿